Springer Series in Statistics
Probability and its Applications

A Series of the Applied Probability Trust

Editors—Probability and its Applications
J. Gani, C.C. Heyde

Editors—Springer Series in Statistics
J. Berger, S. Fienberg, J. Gani, K. Krickeberg,
I. Olkin, B. Singer

Springer Series in Statistics

Andrews/Herzberg: Data: A Collection of Problems from Many Fields for the Student and Research Worker.
Anscombe: Computing in Statistical Science through APL.
Berger: Statistical Decision Theory and Bayesian Analysis, 2nd edition.
Brémaud: Point Processes and Queues: Martingale Dynamics.
Brockwell/Davis: Time Series: Theory and Methods, 2nd edition.
Daley/Vere-Jones: An Introduction to the Theory of Point Processes.
Dzhaparidze: Parameter Estimation and Hypothesis Testing in Spectral Analysis of Stationary Time Series.
Farrell: Multivariate Calculation.
Fienberg/Hoaglin/Kruskal/Tanur (Eds.): A Statistical Model: Frederick Mosteller's Contributions to Statistics, Science, and Public Policy.
Goodman/Kruskal: Measures of Association for Cross Classifications.
Grandell: Aspects of Risk Theory.
Härdle: Smoothing Techniques: With Implementation in S.
Hartigan: Bayes Theory.
Heyer: Theory of Statistical Experiments.
Jolliffe: Principal Component Analysis.
Kres: Statistical Tables for Multivariate Analysis.
Leadbetter/Lindgren/Rootzén: Extremes and Related Properties of Random Sequences and Processes.
Le Cam: Asymptotic Methods in Statistical Decision Theory.
Le Cam/Yang: Asymptotics in Statistics: Some Basic Concepts.
Manoukian: Modern Concepts and Theorems of Mathematical Statistics.
Miller, Jr.: Simultaneous Statistical Inference, 2nd edition.
Mosteller/Wallace: Applied Bayesian and Classical Inference: The Case of *The Federalist Papers*.
Pollard: Convergence of Stochastic Processes.
Pratt/Gibbons: Concepts of Nonparametric Theory.
Read/Cressie: Goodness-of-Fit Statistics for Discrete Multivariate Data.
Reiss: Approximate Distributions of Order Statistics: With Applications to Nonparametric Statistics.
Ross: Nonlinear Estimation.
Sachs: Applied Statistics: A Handbook of Techniques, 2nd edition.
Seneta: Non-Negative Matrices and Markov Chains.
Siegmund: Sequential Analysis: Tests and Confidence Intervals.
Tong: The Multivariate Normal Distribution.
Vapnik: Estimation of Dependences Based on Empirical Data.
West/Harrison: Bayesian Forecasting and Dynamic Models.
Wolter: Introduction to Variance Estimation.
Yaglom: Correlation Theory of Stationary and Related Random Functions I: Basic Results.
Yaglom: Correlation Theory of Stationary and Related Random Functions II: Supplementary Notes and References.

Jan Grandell

Aspects of Risk Theory

Springer-Verlag
New York Berlin Heidelberg London
Paris Tokyo Hong Kong Barcelona

Jan Grandell
Department of Mathematics
The Royal Institute of Technology
100 44 Stockholm
Sweden

Series Editors
J. Gani
Department of Statistics
University of California
Santa Barbara, CA 93106
USA

C. C. Heyde
Department of Statistics
Institute of Advanced Studies
The Australian National University
GPO Box 4, Canberra ACT 2601
Australia

Mathematics Subject Classification 60Gxx, 60G35

Printed on acid-free paper

© 1991 Springer-Verlag New York Inc.
All rights reserved. This work may not be translated or copied in whole or in part without the written permission of the publisher (Springer-Verlag New York, Inc., 175 Fifth Avenue, New York, NY 10010, USA), except for brief excerpts in connection with reviews or scholarly analysis. Use in connection with any form of information storage and retrieval, electronic adaptation, computer software, or by similar or dissimilar methodology now known or hereafter developed is forbidden.
The use of general descriptive names, trade names, trademarks, etc., in this publication, even if the former are not especially identified, is not to be taken as a sign that such names, as understood by the Trade Marks and Merchandise Marks Act, may accordingly be used freely by anyone.

Camera-ready copy provided by the author.
Printed and bound by: Edwards Brothers Incorporated, Ann Arbor, Michigan.
Printed in the United States of America.

9 8 7 6 5 4 3 2 1

ISBN 0-387-97368-0 Springer-Verlag New York Berlin Heidelberg
ISBN 3-540-97368-0 Springer-Verlag Berlin Heidelberg New York

Preface

Collective risk theory, as a part of insurance – or actuarial – mathematics, deals with stochastic models of an insurance business. In such a model the occurrence of the claims is described by a point process and the amounts of money to be paid by the company at each claim by a sequence of random variables. The company recieves a certain amount of premium to cover its liability. The difference between the premium income and the (average) cost for the claims is the "safety loading." The company is furthermore assumed to have a certain initial capital u at its disposal. One important problem in collective risk theory is to investigate the "ruin probability," i.e., the probability that the risk business ever becomes negative.

The simplest model – here called the "classical risk model" – is roughly as follows:

I. The point process is a Poisson process.

II. The costs of the claims are described by independent and identically distributed random variables.

III. The point process and the random variables are independent.

IV. The premiums are described by a constant (and deterministic) rate of income.

The classical risk model can be generalized in many ways.

A. The premiums may depend on the result of the risk business. It is natural to let the safety loading at a time t be "small" if the risk business, at that time, attains a large value and vice versa.

B. Inflation and interest may be included in the model.

C. The occurrence of the claims may be described by a more general point process than the Poisson process.

In the present study we focus exclusively on generalization C. The reason is my personal interest, and not because this is necessarily the most important generalization. Dassios and Embrechts (1989) and Delbaen and Haezendonck (1987) are very readable studies focusing mainly on generalizations A and B. Furthermore, we consider only $\Psi(u)$, i.e., the probability of ruin within *infinite* time. Some remarks on ruin within finite time are, however, given in the appendix.

This is a monograph on certain aspects of risk theory and not a textbook in risk theory. The word "aspects" in the title is almost as informative as the words "risk theory." The reader who wants a textbook is recommended to consult Gerber (1979). That book is a fine introduction to risk theory and almost perfect as a prerequisite for this monograph.

While writing this monograph I have had two potential readers in mind.

The actuary who has a good knowledge of classical risk theory and wants to get acquainted with these kind of generalizations. Anyone with a knowledge of risk theory corresponding to Gerber's book is here regarded as an actuary. For the benefit of the actuary several "inserted surveys" are included.

The probabilist who – resonably simply – wants to get an introduction to modern ruin theory. Parts of the surveys on point processes may also be helpful for some probabilists.

Section 1.1 is devoted to the following four basic results, which go back to the pioneering works by Filip Lundberg and Harald Cramér:

$$\Psi(0) = \frac{1}{1+\rho}, \qquad \text{(I)}$$

where ρ is the "relative" safety loading;

$$\Psi(u) = \frac{1}{1+\rho} e^{-\frac{\rho u}{\mu(1+\rho)}} \qquad \text{(II)}$$

when the claim costs are exponentially distributed with mean μ;

the Cramér-Lundberg approximation

$$\lim_{u \to \infty} e^{Ru} \Psi(u) = C, \qquad \text{(III)}$$

where the Lundberg exponent R is given by a functional equation;

the Lundberg inequality

$$\Psi(u) \leq e^{-Ru}. \qquad \text{(IV)}$$

The Cramér-Lundberg approximation is proved by the aid of a "defective renewal equation" – a technique introduced by William Feller. The Lundberg inequality is proved by a "martingale approach" – introduced by Hans Gerber. Those methods are much simpler than the Wiener-Hopf methods, used by Cramér (1955), albeit the results are less general and less detailed.

Sections 1.2 and 1.3 deal with "practical calculations" of ruin probabilities and estimation of the Lundberg exponent, respectively. These sections lie somewhat outside the main theme of the monograph. They are included since – in my opinion – a discussion related to applications naturally belongs in a presentation of risk theory.

In Chapter 2 the exposition of point processes starts. The chapter may be viewed as an introduction to point processes. It is – hopefully – suited to actuaries.

One main deficiency in the classical risk model is that the possibility of an increase of the insurance business is not taken into account. Generally that possibility is taken care of by introduction of an "operational time scale." In Section 2.1 the martingale approach to point processes is discussed and a "stochastic operational time scale" is defined with the aid of the "compensator of the point process."

The purpose of Section 2.2 is to discuss the choice of the point process describing the occurrence of the claims. That discussion is based on the general theory of point processes. An idea going back to Bertil Almer – one of the Swedish pioneers in risk theory – is taken up, and leads to considerations about thinning of point processes.

A natural – at least from an analytical point of view – generalization of Poisson processes are renewal processes. In Chapter 3 it is shown that (**I**) – (**IV**), essentially, hold also in that case. Chapter 3 has a similar relation to the investigations by Olof Thorin as Section 1.1 has to Cramér (1955).

Another natural generalization of the Poisson process is the Cox process. A Cox process is a generalization in the sense that stochastic variation in the intensity is allowed. Intuitively we shall think of a Cox process N as generated in the following way: first a realization of an intensity process, i.e., a non-negative random process, $\lambda(t)$ is generated and, conditioned upon its realization, N is a non-homogeneous Poisson process with that realization as its intensity. Cox processes are very natural as models for "risk fluctuation." The generalization also seems natural from the discussions in Chapter 2 about thinning of point processes. Chapter 4 is devoted to risk models where the occurrence of the claims is described by a Cox process.

In Section 4.1 analogs to (**I**) and (**II**) are studied when the intensity is markovian. In Section 4.2 the following weaker version of (**IV**) is proved under general assumptions by martingale methods:

there is an $R > 0$ such that for each $\epsilon \in (0, R]$ we have

$$\Psi(u) \leq C_\epsilon e^{-(R-\epsilon)u},$$

where $C_\epsilon < \infty$.

In this generality it is difficult to actually determine R. Furthermore we do not know at all if R is the "right" exponent. These questions are discussed in Sections 4.3 and 4.4 for special intensity processes. In Section 4.5 the unpleasant assumption "$\epsilon > 0$" is removed for certain markovian intensity processes.

In Chapter 5 it is shown that (**I**) holds in great generality.

All definitions, theorems, propositions, examples, and remarks are numbered consecutively within each chapter. Figures and tables have, within each chapter, their own numbering. So, for example, Definition 6 in Section 1.1 is refered to as "Definition 6" in the whole of Chapter 1 and as "Definition 1.6" in the other chapters. The end of proofs is marked by ∎ and the end of examples and remarks is marked by □. The end of the "inserted surveys," which may contain examples and remarks, is marked by □ □.

It is a pleasure to thank Gunnar Englund, Lars Holst, Thomas Höglund, Olof Thorin, Claes Trygger, and Nikos Yannaros for their substantial help in the preparation of this monograph. I am also highly indebted to Tomas Björk who had no objection to my extensive use of our common results and to Gunnar Karlsson who many times helped me with fine details in the typesetting system TeX.

Stockholm, June 1990   Jan Grandell

Contents

Preface	v
1 The classical risk model	**1**
1.1 Ruin probabilities for the classical risk process	4
1.2 "Practical" evaluation of ruin probabilities	13
1.3 Inference for the risk process	25
2 Generalizations of the classical risk model	**33**
2.1 Models allowing for size fluctuation	33
2.2 Models allowing for risk fluctuation	41
3 Renewal models	**57**
3.1 Ordinary renewal models	57
3.2 Stationary renewal models	67
3.3 Numerical illustrations	70
4 Cox models	**77**
4.1 Markovian intensity: Preliminaries	77
4.2 The martingale approach	92
4.3 Independent jump intensity	95
4.3.1 An inbedded random walk	98
4.3.2 Ordinary independent jump intensity	101
4.3.3 Stationary independent jump intensity	103
4.4 Markov renewal intensity	105
4.5 Markovian intensity	112
4.5.1 Application of the basic approach	112
4.5.2 An alternative approach	113
4.6 Numerical illustrations	119
5 Stationary models	**125**

Appendix. Finite time ruin probabilities 135

A.1 The classical model 136
A.2 Renewal models 145
A.3 Cox models 152
A.4 Diffusion approximations 159

References and author index 167

Subject index 173

INSERTED SURVEYS

Basic martingale theory	9 – 10
Basic facts about weak convergence	15 – 16
Point processes and martingales	38 – 40
Point processes and random measures	41 – 53
Basic definitions	41
Superposition of point processes	44
Thinning of point processes	45
Basic Markov process theory	78 – 83
Stationary point processes	128 – 132

CHAPTER 1

The classical risk model

The traditional approach in the *collective risk theory* is to consider a model of the risk business of an insurance company, and to study the probability of ruin, i.e., the probability that the risk business ever will be below some specified (negative) value.

We start with formulating the usual risk model. Let $(\Omega, \mathcal{F}, P)$ be a complete probability space carrying the following* independent objects:

(i)   a point process $N = \{N(t); \ t \geq 0\}$ with $N(0) = 0$;

(ii)   a sequence $\{Z_k\}_1^\infty$ of independent and identically distributed random variables, having the common distribution function F, with $F(0) = 0$, mean value μ, and variance σ^2.

DEFINITION 1. The *risk process*, X, is defined by

$$X(t) = ct - \sum_{k=1}^{N(t)} Z_k, \quad \left(\sum_{k=1}^{0} Z_k \stackrel{\text{def}}{=} 0\right), \tag{1}$$

where c is a positive real constant.

This is the standard model of an insurance company, where $N(t)$ is to be interpreted as the number of claims on the company during the interval $(0, t]$. At each point of N the company has to pay out a stochastic amount of money, and the company receives (deterministically) c units of money per unit time. The constant c is called the *gross risk premium* rate.

Assume that N has intensity α, i.e., $E[N(t)] = \alpha t$. The profit of the risk business over the interval $(0, t]$ is $X(t)$ and thus the expected profit is

$$E[X(t)] = ct - E[N(t)]E[Z_k] = (c - \alpha\mu)t.$$

The relative *safety loading* ρ is defined by

$$\rho = \frac{c - \alpha\mu}{\alpha\mu} = \frac{c}{\alpha\mu} - 1.$$

---

* The probability space may also carry other objects.

The risk process X is said to have *positive* safety loading if $\rho > 0$. Then $X(t)$ has a drift to $+\infty$.

We can now define the *ruin probability* $\Psi(u)$ of a company facing the risk process (1) and having initial capital u.

DEFINITION 2. $\Psi(u) = P\{u + X(t) < 0 \text{ for some } t > 0\}$.

It is sometimes convenient to use the non-ruin probability

$$\Phi(u) = 1 - \Psi(u).$$

Note that it follows from the definition that $\Psi(u) = 1$ for $u < 0$.

In Figure 1 we illustrate the introduced notation, together with some notation to be used later. The "realization" of the risk process is chosen such that the notation shall appear clearly.

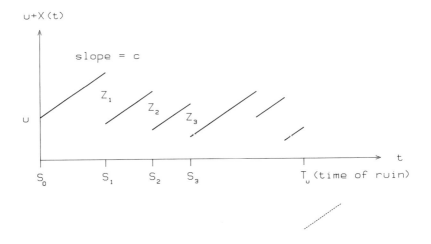

FIGURE 1.   Illustration of notation.

In order to give some idea of a "typical" realization of a risk process we show in Figure 2 a random generation of the "simplest" case. In that case when N is a Poisson process with $\alpha = 1$, the claims are exponentially distributed with $\mu = 1$ and $c = 1.2$. The dotted line represent the mean, i.e., $u + (c - \alpha\mu)t$. We will return to this case in Example 7. Since it is difficult to see any details of the risk process, we give in Figure 3 a magnification of the dotted rectangle.

The function $h(r)$ will be of fundamental importance.

DEFINITION 3. $h(r) = \int_0^\infty e^{rz}\, dF(z) - 1$.

ASSUMPTION 4. *We assume that there exists $r_\infty > 0$ such that $h(r) \uparrow +\infty$ when $r \uparrow r_\infty$ (we allow for the possibility $r_\infty = +\infty$).*

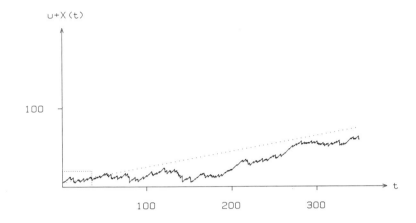

FIGURE 2. Randomly generated risk process.

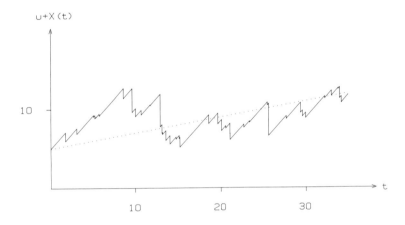

FIGURE 3. Randomly generated risk process.

REMARK 5. It is easily seen that $h(0) = 0$ and that h is increasing, convex and continuous on $[0, r_\infty)$. The important part of Assumption 4 is that $h(r) < \infty$ for some $r > 0$. This means that the tail of dF decreases at least exponentially fast, and thus for example the lognormal and the Pareto distributions are not allowed. Further, the rather pathological case when $h(r_\infty-) < \infty$ and $h(r) = \infty$ for $r > r_\infty$ is excluded. The example $F'(z) = \frac{\text{const.}}{z^2} e^{-z}$ for $z > 1$ shows that such cases do exist.

1.1 Ruin probabilities for the classical risk process

DEFINITION 6. X is called a *classical* risk process or a *Poisson model* if N is a Poisson process.

When nothing else is said, we assume in this section that X is a classical risk process.

A simple way to get an equation for Φ is to use a *"differential"* argument, see, e.g., Cramér (1930, p. 75). Then we consider $X(t)$ in a small time interval $(0, \Delta]$ and separate the four possible cases as follows:

1. no claim occurs in $(0, \Delta]$,
2. one claim occurs in $(0, \Delta]$, but the amount to be paid does not cause ruin,
3. one claim occurs in $(0, \Delta]$, and the amount to be paid does cause ruin, and
4. more than one claim occurs in $(0, \Delta]$.

From the fact that $X(t)$ has stationary and independent increments we get, provided $\Phi(u)$ is differentiable,

$$\Phi(u) = (1 - \alpha\Delta + o(\Delta))\Phi(u + c\Delta) +$$

$$(\alpha\Delta + o(\Delta))\int_0^{u+c\Delta}\Phi(u + c\Delta - z)\,dF(z) + (\alpha\Delta + o(\Delta))\cdot 0 + o(\Delta)$$

$$= (1 - \alpha\Delta)\Phi(u + c\Delta) + \alpha\Delta\int_0^{u+c\Delta}\Phi(u + c\Delta - z)\,dF(z) + o(\Delta)$$

$$= \Phi(u) + c\Delta\Phi'(u) - \alpha\Delta\Phi(u) + \alpha\Delta\int_0^{u}\Phi(u - z)\,dF(z) + o(\Delta), \quad (2)$$

where as usual $o(\Delta)$ means that $o(\Delta)/\Delta \to 0$ as $\Delta \to 0$. Thus we get

$$\Phi'(u) = \frac{\alpha}{c}\Phi(u) - \frac{\alpha}{c}\int_0^{u}\Phi(u - z)\,dF(z). \quad (3)$$

The derivation of (3) is certainly not mathematically satisfying. Although Cramér (1955, pp. 60 - 61) gives a stringent "version" of the differential argument, we shall consider another approach.

Following Feller (1971, p. 183) we shall now derive (3) by a *"renewal"* argument. Let S_1 be the epoch of the first claim. Then we have $X(S_1) = cS_1 - Z_1$. Since the Poisson process is a renewal process and since ruin can not occur in $(0, S_1)$ we have

$$\Phi(u) = E[\Phi(u + cS_1 - Z_1)] = \int_0^{\infty}\alpha e^{-\alpha s}\int_0^{u+cs}\Phi(u + cs - z)\,dF(z)\,ds.$$

1.1 Ruin probabilities for the classical risk process

The change of variables $x = u + cs$ leads to
$$\Phi(u) = \frac{\alpha}{c} e^{\alpha u/c} \int_u^\infty e^{-\alpha x/c} \int_0^x \Phi(x-z)\, dF(z)dx.$$

Consequently Φ is differentiable and differentiation leads to (3). Integrating (3) over $(0, t)$ yields

$$\Phi(t) - \Phi(0) = \frac{\alpha}{c} \int_0^t \Phi(u)\, du + \frac{\alpha}{c} \int_0^t \int_0^u \Phi(u-z) \overset{d}{(1-F(z))}\, du$$

$$= \frac{\alpha}{c} \int_0^t \Phi(u)\, du +$$

$$\frac{\alpha}{c} \int_0^t \left[\Phi(0)(1-F(u)) - \Phi(u) + \int_0^u \Phi'(u-z) d(1-F(z))\, dz \right] du$$

$$= \frac{\alpha}{c} \Phi(0) \int_0^t (1-F(u))\, du + \frac{\alpha}{c} \int_0^t (1-F(z))\, dz \int_z^t \Phi'(u-z)\, du$$

$$= \frac{\alpha}{c} \Phi(0) \int_0^t (1-F(u))\, du + \frac{\alpha}{c} \int_0^t (1-F(z))(\Phi(t-z) - \Phi(0))\, dz.$$

Thus we have
$$\Phi(u) = \Phi(0) + \frac{\alpha}{c} \int_0^u \Phi(u-z)(1-F(z))\, dz. \qquad (4)$$

By monotone convergence it follows from (4), as $u \to \infty$, that
$$\Phi(\infty) = \Phi(0) + \frac{\alpha\mu}{c} \Phi(\infty). \qquad (5)$$

It follows from the law of large numbers that $\lim_{t\to\infty} X(t)/t = c - \alpha\mu$ with probability one. In the case of positive safety loading, $c > \alpha\mu$, there exists a random variable T, i.e., a function of N and $\{Z_k\}$, such that $X(t) > 0$ for all $t > T$. Since only finitely many claims can occur before T it follows that $\inf_{t>0} X(t)$ is finite with probability one and thus $\Phi(\infty) = 1$. Thus $1 = (1 - \Psi(0)) + \frac{\alpha\mu}{c}$ or

$$\Psi(0) = \frac{\alpha\mu}{c} = \frac{1}{1+\rho} \qquad \text{when } c > \alpha\mu. \qquad (I)$$

This is an *insensitivity* or robustness result, since $\Psi(0)$ only depends on ρ and thus on F only through its mean.

EXAMPLE 7. EXPONENTIALLY DISTRIBUTED CLAIMS. Consider the simple case, illustrated in Figures 2 and 3, when Z_k is exponentially distributed. Then (3) is reduced to

$$\Phi'(u) = \frac{\alpha}{c}\Phi(u) - \frac{\alpha}{c\mu} \int_0^u \Phi(u-z) e^{-z/\mu}\, dz$$

$$= \frac{\alpha}{c}\Phi(u) - \frac{\alpha}{c\mu} \int_0^u \Phi(z) e^{-(u-z)/\mu}\, dz.$$

Differentiation leads to

$$\Phi''(u) = \frac{\alpha}{c}\Phi'(u) + \frac{1}{\mu}(\frac{\alpha}{c}\Phi(u) - \Phi'(u)) - \frac{\alpha}{c\mu}\Phi(u)$$

$$= (\frac{\alpha}{c} - \frac{1}{\mu})\Phi'(u) = -\frac{\rho}{\mu(1+\rho)}\Phi'(u)$$

and thus

$$\Phi(u) = C_1 - C_2 e^{-\rho u/(\mu(1+\rho))}.$$

For $\rho > 0$ we have $\Phi(\infty) = 1$ and $\Phi(0) = 1 - \frac{1}{1+\rho}$, which implies $\Phi(u) = 1 - \frac{1}{1+\rho} e^{-\rho u/(\mu(1+\rho))}$ or

$$\Psi(u) = \frac{1}{1+\rho} e^{-\frac{\rho u}{\mu(1+\rho)}}. \tag{II}$$

□

From (4) and (I) we get the following when $c > \alpha\mu$:

$$1 - \Psi(u) = 1 - \frac{\alpha\mu}{c} + \frac{\alpha}{c}\int_0^u (1 - \Psi(u-z))(1 - F(z))\,dz$$

$$= 1 - \frac{\alpha}{c}\left(\mu - \int_0^u (1 - F(z))\,dz + \int_0^u \Psi(u-z)(1 - F(z))\,dz\right)$$

or

$$\Psi(u) = \frac{\alpha}{c}\int_u^\infty (1 - F(z))\,dz + \frac{\alpha}{c}\int_0^u \Psi(u-z)(1 - F(z))\,dz. \tag{6}$$

Since $\int_0^\infty \frac{\alpha}{c}(1 - F(z))\,dz = \frac{\alpha\mu}{c} < 1$ the equation (6) is a defective renewal equation. Following Feller (1971, p. 376) we assume that there exists a constant R such that

$$\frac{\alpha}{c}\int_0^\infty e^{Rz}(1 - F(z))\,dz = 1. \tag{7}$$

Then $\frac{\alpha}{c}e^{Rz}(1 - F(z))$ is the density of a proper probability distribution. Multiplication of (6) by e^{Ru} yields

$$e^{Ru}\Psi(u)$$

$$= \frac{\alpha}{c}e^{Ru}\int_u^\infty (1 - F(z))\,dz + \frac{\alpha}{c}\int_0^u e^{R(u-z)}\Psi(u-z)e^{Rz}(1 - F(z))\,dz \tag{8}$$

which is a proper renewal equation. From the renewal theorem, see Feller (1971, p. 363), it then follows that

$$\lim_{u\to\infty} e^{Ru}\Psi(u) = \frac{C_1}{C_2}, \tag{9}$$

where

$$C_1 = \frac{\alpha}{c}\int_0^\infty e^{Ru}\int_u^\infty (1 - F(z))\,dz\,du \tag{10}$$

1.1 Ruin probabilities for the classical risk process

and
$$C_2 = \frac{\alpha}{c} \int_0^\infty z e^{Rz}(1 - F(z))\, dz \tag{11}$$

provided R, C_1, and C_2 exist in $(0, \infty)$. We shall now, for the first time, rely on the important Assumption 4. We get from (7),

$$\frac{c}{\alpha} = \int_0^\infty e^{Rz}(1 - F(z))\, dz = -\frac{1}{R} + \frac{1}{R}\int_0^\infty e^{Rz}\, dF(z) = \frac{h(R)}{R}$$

and thus (the "*Lundberg exponent*") R is the positive solution of

$$h(r) = \frac{cr}{\alpha}. \tag{12}$$

Further, we have

$$C_1 = \frac{\alpha}{c}\int_0^\infty e^{Ru}\int_u^\infty (1 - F(z))\, dz\, du$$

$$= -\frac{\alpha}{cR}\int_0^\infty (1 - F(z))\, dz + \frac{\alpha}{c}\int_0^\infty \frac{1}{R}e^{Rz}(1 - F(z))\, dz$$

$$= \frac{1}{R}(1 - \frac{\alpha\mu}{c}) = \frac{1}{R}\frac{\rho}{1+\rho}.$$

Using $h'(R) = \int_0^\infty z e^{Rz}\, dF(z)$ and $\int z e^{Rz}\, dz = (\frac{z}{R} - \frac{1}{R^2})e^{Rz}$ we get

$$C_2 = \frac{\alpha}{c}\int_0^\infty z e^{Rz}(1 - F(z))\, dz = \frac{\alpha}{c}\left(\frac{1}{R^2} + \int_0^\infty (\frac{z}{R} - \frac{1}{R^2})e^{Rz}\, dF(z)\right)$$

$$= \frac{\alpha}{c}\left(\frac{1}{R^2} + \frac{h'(R)}{R} - \frac{h(R)+1}{R^2}\right) = \frac{\alpha}{c}\left(\frac{h'(R)}{R} - \frac{c}{\alpha R}\right)$$

$$= \frac{\alpha\mu}{c}\frac{1}{R}\frac{1}{\mu}(h'(R) - c/\alpha) = \frac{1}{1+\rho}\frac{1}{R}\frac{1}{\mu}(h'(R) - c/\alpha).$$

Thus we have
$$\lim_{u \to \infty} e^{Ru}\Psi(u) = \frac{\rho\mu}{h'(R) - c/\alpha} \tag{III}$$

which is called the "*Cramér-Lundberg approximation.*"

EXAMPLE 7. CONTINUED. We have $h(r) = \frac{1}{\mu}\int_0^\infty e^{rz}e^{-z/\mu}dz - 1 = \frac{\mu r}{1-\mu r}$ and thus R is the positive solution of $\frac{\mu r}{1-\mu r} = \frac{cr}{\alpha}$, i.e.,

$$R = \frac{\rho}{\mu(1+\rho)}.$$

Further, $h'(R) = \frac{\mu}{(1-\mu R)^2} = \mu(1+\rho)^2$ and thus

$$\frac{\rho\mu}{h'(R) - c/\alpha} = \frac{\rho\mu}{\mu(1+\rho)^2 - c/\alpha} = \frac{\rho}{(1+\rho)^2 - (1+\rho)} = \frac{1}{1+\rho}.$$

As is to be expected, the Cramér-Lundberg approximation is exact when

8   1 The classical risk model

the claims are exponentially distributed.
□

EXAMPLE 8. LIFE ANNUITY INSURANCE. We have assumed that $c > 0$ and that $F(0) = 0$. In insurance terminology this is generally called *positive risk sums*, and includes most non-life branches and also the ordinary types of life insurance, where a certain amount of money is paid at the death of a policyholder.

There are, however, situations where the circumstances are reversed, i.e., where $c < 0$ and $F(0) = 1$. The typical example is life annuity, or pension, insurance, where a life annuity rate $-c$ is paid from the company to the policyholder and where the claim, i.e., the death of the policyholder, will place an amount of money corresponding to the "expected pension to be paid" at the company's free disposal. Thus the claim means an income, or a negative cost, for the company. This situation is generally called *negative risk sums*.

Thus, in this example, we assume that $c < 0$ and $F(0) = 1$. Thus ruin can only occur between claims. If ruin occurs, the risk process must at some epoch take the value x, $0 \leq x \leq u$, for the first time. Since the risk process has stationary and independent increments it follows that

$$\Psi(u) = \Psi(u - x)\Psi(x) \qquad \text{for } 0 \leq x \leq u \tag{13}$$

and thus $\Psi(u) = e^{-ru}$ for some value of r. Obviously the "renewal argument" does not work, but the obvious modifications of the "differential argument" leads to

$$\Phi'(u) = \frac{\alpha}{c}\Phi(u) - \frac{\alpha}{c}\int_{-\infty}^{0} \Phi(u - z)\, dF(z), \tag{14}$$

or, if we use $\Phi(u) = 1 - e^{-ru}$,

$$re^{-ru} = \frac{\alpha}{c}\left(1 - e^{-ru} - 1 + \int_{-\infty}^{0} e^{-r(u-z)}\, dF(z)\right)$$

and thus

$$\frac{cr}{\alpha} = \int_{-\infty}^{0} e^{rz}\, dF(z) - 1. \tag{15}$$

Comparing Definition 3 and (12) we realize that (15) is the natural definition of Lundberg's exponent in this case and that

$$\Psi(u) = e^{-Ru}. \tag{16}$$

□

We shall now consider a completely different approach, due to Gerber (1973), which uses martingales. Before considering his approach, we shall need some basic facts about martingales.

1.1 Ruin probabilities for the classical risk process

BASIC MARTINGALE THEORY

The definitions and results to be given here can be found in any standard text on martingales, like Elliott (1982).

We shall, for future purposes, be somewhat more general than is really needed for the moment. Therefore the probability space $(\Omega, \mathcal{F}, P)$ may carry more objects than the risk process.

DEFINITION 9. A *filtration* $\mathbf{F} = (\mathcal{F}_t;\ t \geq 0)$ is a non-decreasing family of sub-σ-algebras of $\mathcal{F}$.

DEFINITION 10. Let for any process $Y = \{Y(t);\ t \geq 0\}$, the filtration $\mathbf{F}^Y = (\mathcal{F}_t^Y;\ t \geq 0)$ be defined by

$$\mathcal{F}_t^Y = \sigma\{Y(s);\ s \leq t\}.$$

Thus $\mathcal{F}_t^Y$ is the σ-algebra generated by Y up to time t, and represents the *history* of Y up to time t. Y is *adapted* to $\mathbf{F}$, i.e., Y is $\mathcal{F}_t$-measurable for all $t \geq 0$, if and only if $\mathcal{F}_t^Y \subseteq \mathcal{F}_t$ for all $t \geq 0$.

DEFINITION 11. An $\mathbf{F}$-*martingale* (respectively $\mathbf{F}$-*supermartingale*)

$$M = \{M(t);\ t \geq 0\}$$

is a real valued process such that:

(i) $M(t)$ is $\mathcal{F}_t$-measurable for $t \geq 0$;

(ii) $E[|M(t)|] < \infty$ for $t \geq 0$;

(iii) $E^{\mathcal{F}_s}[M(t)] = E[M(t) \mid \mathcal{F}_s] = (\leq) M(s)$ P-a.s. for $t \geq s$.

DEFINITION 12. An $\mathbf{F}$-martingale or an $\mathbf{F}$-supermartingale M is called *right continuous* if as follows:

(i)   the trajectories $M(t)$ are right continuous;

(ii)   the filtration $\mathbf{F}$ is right continuous, i.e.,

$$\mathcal{F}_t = \bigcap_{s>t} \mathcal{F}_s \qquad \text{for } t \geq 0.$$

All processes which we consider have right continuous trajectories and the filtrations are so simple that the condition of right continuity is of no problem.

DEFINITION 13. A random variable T, $\Omega \to [0, \infty]$, is an $\mathbf{F}$-*stopping* time if $\{T \leq t\} \in \mathcal{F}_t$ for each $t \geq 0$.

This means that, knowing the history up to time t, one can decide if $T \leq t$ or not. Note that the outcome $T = \infty$ is allowed. If T is a stopping time, so is $t \wedge T = \min(t, T)$ for each t.

The following simplified version of the "Optional Stopping Theorem" is essential for our applications.

THEOREM 14. *Let T be a bounded stopping time, i.e., $T \leq t_0 < \infty$, and M a right continuous **F**-martingale (**F**-supermartingale). Then*

$$E^{\mathcal{F}_0}[M(T)] = (\leq) \ M(0) \qquad P\text{-a.s.}$$

□ □

Now we consider the *"martingale approach."* Let $Y(t)$ be a right continuous process such that:

(i) $\quad Y(0) = 0 \quad$ P-a.s.;

(ii) $\quad Y$ has stationary and independent increments;

(iii) $\quad E[Y(t)] = t\beta, \qquad$ where $\beta > 0$;

(iv) $\quad E[e^{-rY(t)}] < \infty$ for some $r > 0$.

Then
$$E[e^{-rY(t)}] = e^{tg(r)} \qquad \text{for some function } g(\cdot).$$

If Y is a classical risk process with positive safety loading we have $\beta = c - \alpha\mu$. Further, we have

$$E[e^{-rY(t)}] = e^{-rct} \sum_{k=0}^{\infty} \frac{(\alpha t)^k}{k!} e^{-\alpha t}(h(r)+1)^k$$

$$= e^{-rct+\alpha t(h(r)+1)-\alpha t} = e^{t(\alpha h(r)-rc)}$$

and thus $g(r) = \alpha h(r) - rc$.

Note that Y may also, for example, be the risk process corresponding to life annuity insurance or the Wiener process with positive drift.

Let T_u be the time of ruin, i.e.,

$$T_u = \inf\{t \geq 0 \mid u + Y(t) < 0\}.$$

Obviously T_u is a $\mathbf{F}^Y$-stopping time and note that $\Psi(u) = P\{T_u < \infty\}$. Put
$$M_u(t) = \frac{e^{-r(u+Y(t))}}{e^{tg(r)}}.$$

M_u is an $\mathbf{F}^Y$-martingale, since

$$E^{\mathcal{F}^Y_s}[M_u(t)] = E^{\mathcal{F}^Y_s}\left[\frac{e^{-r(u+Y(t))}}{e^{tg(r)}}\right]$$

$$= E^{\mathcal{F}^Y_s}\left[\frac{e^{-r(u+Y(s))}}{e^{sg(r)}} \cdot \frac{e^{-r(Y(t)-Y(s))}}{e^{(t-s)g(r)}}\right]$$

$$= M_u(s) \cdot E^{\mathcal{F}^Y_s}\left[\frac{e^{-r(Y(t)-Y(s))}}{e^{(t-s)g(r)}}\right] = M_u(s). \qquad (17)$$

1.1 Ruin probabilities for the classical risk process

Choose $t_0 < \infty$ and consider $t_0 \wedge T_u$ which is a bounded $\mathbf{F}^Y$-stopping time. Since $\mathcal{F}_0^Y$ is trivial and since M_u is positive, it follows from Theorem 14 that

$$e^{-ru} = M_u(0) = E[M_u(t_0 \wedge T_u)]$$
$$= E[M_u(t_0 \wedge T_u) \mid T_u \leq t_0]P\{T_u \leq t_0\} +$$
$$\quad E[M_u(t_0 \wedge T_u) \mid T_u > t_0]P\{T_u > t_0\}$$
$$\geq E[M_u(t_0 \wedge T_u) \mid T_u \leq t_0]P\{T_u \leq t_0\}$$
$$= E[M_u(T_u) \mid T_u \leq t_0]P\{T_u \leq t_0\} \tag{18}$$

and thus, since $u + Y(T_u) \leq 0$ on $\{T_u < \infty\}$,

$$P\{T_u \leq t_0\} \leq \frac{e^{-ru}}{E[M_u(T_u) \mid T_u \leq t_0]} \leq \frac{e^{-ru}}{E[e^{-T_u g(r)} \mid T_u \leq t_0]}$$
$$\leq e^{-ru} \sup_{0 \leq t \leq t_0} e^{tg(r)}. \tag{19}$$

Let $t_0 \to \infty$ in (19). Then we get

$$\Psi(u) \leq e^{-ru} \sup_{t \geq 0} e^{tg(r)}. \tag{20}$$

In order to get this inequality as good as possible, we shall choose r as large as possible under the restriction $\sup_{t \geq 0} e^{tg(r)} < \infty$. Let R denote that value. Obviously this means that

$$R = \sup\{r \mid g(r) \leq 0\}. \tag{21}$$

In the classical risk process case this gives R as the positive solution of $h(r) = cr/\alpha$, i.e., R is the Lundberg exponent. Thus we have

$$\Psi(u) \leq e^{-Ru} \tag{IV}$$

which is called the "*Lundberg inequality.*" Comparing with (**III**) and (**IV**) it is seen that R actually is the best possible exponent.

REMARK 15. We shall indicate a slight variation of the derivation given in (17) - (21). Put $M_u(t) = e^{-r(u+Y(t))}$. Then, see (17),

$$E^{\mathcal{F}_s}[M_u(t)] = M_u(s) \cdot e^{(t-s)g(r)}$$

and thus M_u is an $\mathbf{F}^Y$-supermartingale if and only if $g(r) \leq 0$. Equation (18) holds for $\mathbf{F}^Y$-supermartingales, except that $M_u(0) \geq E[M_u(t_0 \wedge T_u)]$. Thus we get $\Psi(u) \leq e^{-ru}$, where r shall be chosen as large as possible under the restriction that M_u is an $\mathbf{F}^Y$-supermartingale. Obviously this leads to (21).
□

Let us go back to (18). Then we have, with $r = R$,

$$e^{-Ru} = E[e^{-R(u+Y(T_u))} \mid T_u \leq t_0]P\{T_u \leq t_0\} +$$
$$\quad E[e^{-R(u+Y(t_0))} \mid T_u > t_0]P\{T_u > t_0\}. \tag{22}$$

Let $I\{A\}$ denote the indicator function for the set A. Then we have
$$0 \leq E[e^{-R(u+Y(t_0))} \mid T_u > t_0] P\{T_u > t_0\}$$
$$= E[e^{-R(u+Y(t_0))} I\{T_u > t_0\}] \leq E[e^{-R(u+Y(t_0))} I\{u + Y(t_0) \geq 0\}].$$
Since $0 \leq e^{-R(u+Y(t_0))} I\{u + Y(t_0) \geq 0\} \leq 1$ it follows, due to the drift of $Y(t)$ to $+\infty$, by dominated convergence that
$$\lim_{t_0 \to \infty} E[e^{-R(u+Y(t_0))} \mid T_u > t_0] P\{T_u > t_0\} = 0$$
and thus we get from (22) that
$$\Psi(u) = \frac{e^{-Ru}}{E[e^{-R(u+Y(T_*))} \mid T_u < \infty]}. \tag{23}$$

EXAMPLE 8. CONTINUED. When $Y(t)$ is continuous at the time of ruin we have $u + Y(T_u) = 0$ on $\{T_u < \infty\}$ and thus $\Psi(u) = e^{-Ru}$. This holds for the risk process corresponding to life annuity insurance and for the Wiener process. Thus we have proved (16) again.
□

EXAMPLE 7. CONTINUED. In this example Z_k is exponentially distributed. The exponential distribution is characterized by its "lack of memory," i.e., that
$$P\{Z_k > z + x \mid Z_k > x\} = P\{Z_k > z\}.$$
Assume that $T_u < \infty$. Put $\mathcal{Z} = -(X(T_u) - X(T_u-))$, i.e., $\mathcal{Z}$ is the cost for that claim which "caused" ruin. Let $\mathcal{F}^X_{T_u-}$ be the σ-algebra "*strictly prior*" to ruin. Formally
$$\mathcal{F}^X_{T_u-} = \sigma\{A \cap \{t < T_u\};\ A \in \mathcal{F}^X_t,\ t \geq 0\}.$$
Intuitively $\mathcal{F}^X_{T_u-}$ is the history of X up to a time immediately before ruin, including *knowledge* of the ruin but *not* of the value of the risk process immediately after ruin. Then we have
$$E[e^{-R(u+X(T_*))} \mid T_u < \infty]$$
$$= E[E^{\mathcal{F}^X_{T_u-}}\{e^{-R(u+X(T_u-)-\mathcal{Z})} \mid T_u < \infty\} \mid T_u < \infty]$$
$$= E[E\{e^{-R(u+X(T_u-)-\mathcal{Z})} \mid X(T_u-), \mathcal{Z} > u + X(T_u-), T_u < \infty\} \mid T_u < \infty]$$
$$= E[h(R) + 1],$$
where the last equality is due to the lack of memory of the exponential distribution. Now
$$R = \frac{\rho}{\mu(1+\rho)} \quad \text{and} \quad h(R) + 1 = \frac{cR}{\alpha} + 1 = \frac{c\rho}{\alpha\mu(1+\rho)} + 1 = \rho + 1$$
and we have a new proof of (II).
□

The martingale approach is very powerful for proving inequalities but it seems, for example, difficult to prove the Cramér-Lundberg approximation.

Formally it is no problem to calculate the ruin probability exactly by using Laplace transforms. Put

$$\hat{f}(v) \stackrel{\text{def}}{=} \int_0^\infty e^{-vz}\, dF(z) = h(-v) + 1 \quad \text{and} \quad \hat{\phi}(v) \stackrel{\text{def}}{=} \int_{0-}^\infty e^{-vz}\, d\Phi(z).$$

From (4) and (**I**) we immediately get

$$\hat{\phi}(v) = 1 - \frac{\alpha\mu}{c} + \hat{\phi}(v)\frac{1-\hat{f}(v)}{v}$$

and thus

$$\hat{\phi}(v) = \frac{1 - \frac{\alpha\mu}{c}}{1 - \frac{\alpha}{c}\frac{1-\hat{f}(v)}{v}} \tag{24}$$

and thus the problem of calculating ruin probabilities is transformed to a problem of inverting Laplace transforms.

NOTES: The title of this section really ought to be "Classical results about ruin probabilities for the classical risk process" since the main results (**I**) - (**IV**) are due to Lundberg (1926) and Cramér (1930), while the general ideas underlying the collective risk theory go back as far as to Lundberg (1903). These works appeared before the theory of stochastic processes was developed and are therefore not quite mathematically stringent. They are pioneering works, not only in risk theory, but also in the development of the general theory of stochastic processes. The development of risk theory using rigorous methods is to a large extent due to Arfwedson, Cramér, Saxén, Segerdahl, and Täcklind. For a survey of their contribution we refer to Cramér (1955, pp. 48 - 51), where a stringent presentation of risk theory is given. Cramér's analysis is based on Wiener-Hopf methods applied to the integral equation (6) and generalizations of it.

1.2 "Practical" evaluation of ruin probabilities

In our presentation of risk theory we have only given an explicit form of the ruin probability when the claims are exponentially distributed and occur according to a Poisson process. There is really no reason to believe that the exponential distribution is a realistic description of the claim behavior. In many branches of non-life insurance the exponential distribution is regarded as much to "kind," and even Assumption 4 is regarded as being too restrictive. There are several cases of claim distributions where $\Psi(u)$ can be explicitly given or, at least, given on a form suited for numerical calculation.

The simplest possible example is probably when the claim distribution is a finite mixture of exponential distributions, i.e.,

$$F(z) = \sum_{k=1}^{n}(1 - e^{-z\theta_k})p_k,$$

where $p_k \geq 0$, $\sum_{k=1}^{n} p_k = 1$, and $\theta_k \ (= 1/\mu_k) \geq 0$. Then the trick used in Example 7 to eliminate the integral works, see Gerber (1979, p. 117). The trick has, however, practical relevance only for very small values of n. Thorin (1973) considered the much more general case

$$F(z) = \int_0^\infty (1 - e^{-z\theta})\, dV(\theta), \tag{25}$$

where V is a distribution function with $V(0) = 0$. He obtained a formula for $\Psi(u)$ which requires no transform inversion. For suitable choices of V, the Pareto distribution (Seal, 1969) and some Γ-distributions (Thorin, 1973) have the representation (25). If $V'(\theta)$ is allowed to change sign, the lognormal distribution (Thorin and Wikstad, 1977) also has the representation (25). For a survey of numerical calculations of ruin probabilities we refer to Thorin (1977).

EXAMPLE 16. As an illustration of the results obtained by Thorin (1973) we consider the Γ-distribution. For

$$V'(\theta) = \frac{1}{\pi}(\sin\frac{\pi}{\beta})\theta^{-1}(\beta\theta - 1)^{-1/\beta} \qquad \theta \geq \frac{1}{\beta} \quad \text{and} \quad \beta > 1$$

one gets

$$F'(z) = \frac{z^{(1/\beta)-1}}{\beta^{1/\beta}\,\Gamma(1/\beta)} e^{-z/\beta} \qquad z > 0$$

which is a Γ-distribution with $\mu = 1$ and $\sigma^2 = \beta$. Then we have (Thorin 1973, p. 118)

$$\Psi(u) = \frac{\rho(1 - \beta R)e^{-Ru}}{1 + (1+\rho)(R + \beta R - 1)} +$$

$$\frac{\rho}{\pi\beta}\sin\frac{\pi}{\beta}\int_0^\infty \frac{x^{1/\beta}e^{-(x+1)u/\beta}\,dx}{\left\{x^{1/\beta}\left[1 + (1+\rho)\frac{x+1}{\beta}\right] - \cos\frac{\pi}{\beta}\right\}^2 + \sin^2\frac{\pi}{\beta}}, \tag{26}$$

where R is the positive solution of $(1 - \beta r)^{-1/\beta} - 1 = (1+\rho)r$ for $r < 1/\beta$.

Note that the first term in (26) is the Cramér-Lundberg approximation. The second term can be used in order to obtain an upper bound for the error in the Cramér-Lundberg approximation. From Grandell and Segerdahl (1971, p. 147) it follows that for $\rho = 0.1$ the error is less than 10^{-6} as soon as $u > 7.26 \cdot \beta$.

For $\beta < 1$ the Γ-distribution F does not have the representation (25). Thorin (1986) has given an expression for $\Psi(u)$, analogous to (26), also in this case. That expression is – in general – an extension of (26).
□

An alternative to numerical calculations is, of course, simulation. The straightforward simulation of $\Psi(u)$, by running N replicates of $X(t)$ and calculating the fraction of runs with ruin will, in general, require an enormous number of random numbers. For a much more promising method of simulation of ruin probabilities we refer to Asmussen (1985).

It is very natural to try to find "simple" and "good" *approximations* of $\Psi(u)$. Several approximations have been proposed. Some of them are more or less ad hoc and their merits can *only* be judged by numerical comparison. Others are based on limit theorems, and the limit procedure may give hints on their applicability. In that case numerical comparison may be needed in order to get information about the speed of convergence.

The most famous approximation is, of course, the Cramér-Lundberg approximation which is good for large values of u and therefore small values of $\Psi(u)$. Practically it is somewhat difficult to apply, since it requires full knowledge of the claim distribution.

The word "ruin" may sound very "dramatic," and one may imagine "old widows starving because they don't get their pension." Certainly it is more realistic to interpret "ruin" as a technical term meaning that some not too dramatic economical procedure must be done in the insurance company. Therefore it may be interesting to look for approximations which work for less small values of $\Psi(u)$. One way to express this is, if the Cramér-Lundberg approximation is regarded to be related to "large deviations," to look for approximations related to "the central limit theorem." Therefore we shall consider *diffusion approximations* where the idea is to approximate the risk process with a Wiener process with drift. Mathematically such approximations are based on the theory of weak convergence of probability measures. Standard references well suited for our applications are Billingsley (1968) and Lindvall (1973).

BASIC FACTS ABOUT WEAK CONVERGENCE

Let D be the space of functions on $[0, \infty)$ that are right-continuous and have left-hand limits. Endowed with the Skorohod J_1 topology, D is a *Polish space*, i.e., separable and metrizable with a complete metric. A stochastic process $X = \{X(t); t \geq 0\}$ is said to be in D if all its realizations are in D. The distribution of X is a probability measure P on D. Let $X, X_1, X_2, \ldots$ be processes in D. We say that X_n converges *in distribution* to X, and we write $X_n \xrightarrow{d} X$, if $E[f(X)] \to E[f(X)]$ for all bounded and continuous real-valued functions f on D. Convergence in distribution of X_n to X implies, for example, and that $\inf_{0 \leq t \leq t_0} X_n(t) \xrightarrow{d} \inf_{0 \leq t \leq t_0} X(t)$ for any $t_0 < \infty$, but it does generally *not* imply $\inf_{t \geq 0} X_n(t) \xrightarrow{d} \inf_{t \geq 0} X(t)$, see Remark 13 in the appendix.

Further, let W be a standard Wiener process, i.e., $W(0) = 0$; $W(t)$ has independent and normally distributed increments such that
$$E[W(t) - W(s)] = 0 \quad \text{and} \quad \text{Var}[W(t) - W(s)] = t - s$$
for $t > s$ and its realizations are continuous.

□ □

To our knowledge the first application of weak convergence in risk theory is due to Iglehart (1969). Our presentation is based on Grandell (1977).

Put $\bar{S}(t) = \sum_{k=1}^{N(t)} Z_k$. It is easy to show that
$$E[\bar{S}(t)] = \alpha\mu t \quad \text{and} \quad \text{Var}[\bar{S}(t)] = \alpha(\mu^2 + \sigma^2)t.$$
Define $\bar{S}_n$ by
$$\bar{S}_n(t) = \frac{\bar{S}(nt) - \alpha\mu nt}{\sqrt{\alpha(\mu^2 + \sigma^2)\,n}}.$$
It is well-known that $\bar{S}_n \xrightarrow{d} W$ as $n \to \infty$. Note that $X(t) = ct - \bar{S}(t)$ and that $\Psi(u) = P\{\inf_{t\geq 0} X(t) < -u\}$. Define Y_n by
$$Y_n(t) = \frac{c_n nt - \bar{S}(nt)}{\sqrt{n}},$$
which means that we let the gross risk premium rate depend on n, and Y by
$$Y(t) = \gamma\alpha\mu t - \sqrt{\alpha(\mu^2 + \sigma^2)}\,W(t)$$
and put $\rho_n = \frac{c_n - \alpha\mu}{\alpha\mu}$. Thus Y is a Wiener process with drift. Since
$$Y_n(t) = \frac{c_n nt - \alpha\mu nt}{\sqrt{n}} - \sqrt{\alpha(\mu^2 + \sigma^2)}\,\bar{S}_n(t)$$
$$= \rho_n \alpha\mu\sqrt{n}\,t - \sqrt{\alpha(\mu^2 + \sigma^2)}\,\bar{S}_n(t)$$
it follows (Grandell 1977, p. 52) that $Y_n \xrightarrow{d} Y$ as $n \to \infty$ if and only if $\rho_n \sqrt{n} \to \gamma$. It also follows from Grandell (1978) that $\inf_{t\geq 0} Y_n(t) \xrightarrow{d} \inf_{t\geq 0} Y(t)$ and thus
$$P\{\inf_{t\geq 0} Y_n(t) < -y\} \to P\{\inf_{t\geq 0} Y(t) < -y\}.$$
Obviously $P\{\inf_{t\geq 0} Y_n(t) < -y\} = \Psi(y\sqrt{n})$, with relative safety loading ρ_n. Further, we have
$$E[e^{-rY(t)}] = e^{tg(r)} = e^{t[-\gamma\alpha\mu r + \alpha(\mu^2 + \sigma^2)\frac{r^2}{2}]}$$
and thus, cf. (21) and Example 8 continued, we get the well-known result
$$P\{\inf_{t\geq 0} Y(t) < -y\} = e^{-y\gamma\frac{2\mu}{\mu^2+\sigma^2}}.$$

This leads to the diffusion approximation

$$\Psi(u) \approx \Psi_D(u) = e^{-u\rho\frac{2\mu}{\mu^2+\sigma^2}} \qquad (27)$$

if ρ is small and u is large in such a way that u and ρ^{-1} are of the same order. In queuing theory (27) is known as the "heavy traffic approximation." The relation between risk theory and queueing theory will be discussed in Remark 5.1.

REMARK 17. When Assumption 4 holds, one can make a comparison with the Lundberg inequality (**IV**). Then we have

$$\frac{cR}{\alpha} = h(R) \geq \mu R + \frac{\mu^2 + \sigma^2}{2} R^2 \quad \text{or} \quad R \leq \frac{2(c/\alpha - \mu)}{\mu^2 + \sigma^2} = \frac{2\rho\mu}{\mu^2 + \sigma^2}$$

and thus $\Psi_D(u) \leq e^{-Ru}$. Thus Lundberg's inequality also holds for Ψ_D. Although this means that Ψ_D can not overestimate Ψ too much it is really a negative result, since Ψ_D may very well – and this is more serious – underestimate Ψ.

□

EXAMPLE 18. In order to get some preliminary apprehension of the behavior of Ψ_D we consider exponentially distributed claims. For the exponential distribution we have $\sigma^2 = \mu^2$. From this and from (**II**) we get

$$\Psi_D(u) = e^{-u\rho/\mu} \quad \text{and} \quad \Psi(u) = \frac{1}{1+\rho} e^{-\frac{u\rho}{\mu(1+\rho)}}.$$

Thus the *relative error*, which is denoted by $\mathcal{E}_D(u)$, is

$$\mathcal{E}_D(u) = \frac{\Psi_D(u) - \Psi(u)}{\Psi(u)} = (1+\rho)e^{-\frac{u\rho^2}{\mu(1+\rho)}} - 1. \qquad (28)$$

Obviously $\mathcal{E}_D(u) \leq \rho$ so, as shown in Remark 17, the "overestimation" is not a serious problem. For specified values of ρ and of $|\mathcal{E}_D(u)|$ we get from (28) that the "allowed" interval of $\frac{u}{\mu}$ is

$$\left(\max(0, \frac{1+\rho}{\rho^2} \log \frac{1+\rho}{1+|\mathcal{E}_D|}), \frac{1+\rho}{\rho^2} \log \frac{1+\rho}{1-|\mathcal{E}_D|}\right) \qquad (29)$$

which, for example, means that

$$\begin{aligned}
\rho &= 10\% \text{ and } |\mathcal{E}_D| = 10\% \text{ give} & (0 \text{ , } 22.07), \\
\rho &= 10\% \text{ and } |\mathcal{E}_D| = 5\% \text{ give} & (5.12 \text{ , } 16.13), \\
\rho &= 10\% \text{ and } |\mathcal{E}_D| = 1\% \text{ give} & (9.39 \text{ , } 11.59), \\
\rho &= 5\% \text{ and } |\mathcal{E}_D| = 10\% \text{ give} & (0 \text{ , } 64.74), \\
\rho &= 5\% \text{ and } |\mathcal{E}_D| = 5\% \text{ give} & (0 \text{ , } 42.04), \\
\rho &= 5\% \text{ and } |\mathcal{E}_D| = 1\% \text{ give} & (16.31 \text{ , } 24.71).
\end{aligned}$$

□

18   1 The classical risk model

Now we consider two proposed approximations, which are *not* based on limit theorems or other theoretical considerations.

The *Beekman-Bowers approximation* is a modification of an approximation proposed by Beekman (1969). The modification is proposed by Bowers in the discussion of Beekman's paper.

Put $H(u) = P\{\inf_{t\geq 0} X(t) < -u \mid \inf_{t\geq 0} X(t) < 0\}$. It follows from (I) that

$$H(u) = \frac{\Phi(u) - \Phi(0)}{1 - \Phi(0)} = 1 - (1+\rho)\Psi(u)$$

or

$$\Psi(u) = \frac{1}{1+\rho}(1 - H(u)).$$

Let μ_H and σ_H^2 denote the mean and the variance corresponding to H. The idea is to replace $H(u)$ with a Γ-distribution function $G(u)$, such that the two first moments of H and G coincide. Then we have

$$\Psi_{BB}(u) = \frac{1}{1+\rho}(1 - G(u)). \tag{30}$$

Using (24) these moments can easily be calculated. Put

$$\zeta_k = E[Z_j^k], \quad k = 1, 2, 3$$

and note that $\zeta_1 = \mu$ and $\zeta_2 = \mu^2 + \sigma^2$. We have

$$\hat{h}(v) \stackrel{\text{def}}{=} \int_0^\infty e^{-vu}\, dH(u) = \frac{\hat{\phi}(v) - 1 - \frac{\alpha\mu}{c}}{\frac{\alpha\mu}{c}} = \frac{c}{\alpha\mu}\hat{\phi}(v) - \rho$$

$$= \frac{\rho}{1 - \frac{1}{1+\rho}\frac{1-\hat{f}(v)}{\zeta_1 v}} - \rho = \frac{\rho(1+\rho)}{1+\rho - \left(1 - \frac{\zeta_2}{2\zeta_1}v + \frac{\zeta_3}{6\zeta_1}v^2 + O(v^3)\right)} - \rho$$

$$= 1 - \frac{\zeta_2(1+\rho)}{2\rho\zeta_1}v + (1+\rho)\left(\frac{\zeta_3}{3\rho\zeta_1} + \frac{\zeta_2^2}{2\rho^2\zeta_1^2}\right)\frac{v^2}{2} + O(v^3)$$

and thus

$$\mu_H = \frac{\zeta_2(1+\rho)}{2\rho\zeta_1} \quad \text{and} \quad \sigma_H^2 = \frac{\zeta_2(1+\rho)}{2\rho\zeta_1}\left(\frac{2\zeta_3}{3\zeta_2} + \frac{\zeta_2(1-\rho)}{2\rho\zeta_1}\right).$$

It follows from the construction that $\Psi_{BB}(0) = \Psi(0)$.

EXAMPLE 18. CONTINUED. For exponentially distributed claims we have $\zeta_1 = \mu$, $\zeta_2 = 2\mu^2$, and $\zeta_3 = 6\mu^3$. Thus

$$\mu_H = \frac{2\mu^2(1+\rho)}{2\rho\mu} = \frac{\mu(1+\rho)}{\rho}$$

and

$$\sigma_H^2 = \frac{\mu(1+\rho)}{\rho}\left(\frac{2}{3}\frac{6\mu^3}{2\mu^2} + \frac{2\mu^2(1-\rho)}{2\rho\mu}\right)$$

$$= \frac{\mu(1+\rho)}{\rho^2}(2\mu\rho + \mu(1-\rho)) = \frac{\mu^2(1+\rho)^2}{\rho^2} = \mu_H^2.$$

A Γ-distribution with these moments is an exponential distribution, and thus

$$\Psi_{BB}(u) = \frac{1}{1+\rho} e^{-\frac{u\rho}{\mu(1+\rho)}} = \Psi(u).$$

□

The *De Vylder approximation*, proposed by De Vylder (1978), is based on the simple, but ingenious, idea to replace the risk process X with a risk process $\tilde{X}$ with exponentially distributed claims such that

$$E[X^k(t)] = E[\tilde{X}^k(t)] \quad \text{for } k = 1, 2, 3.$$

$\tilde{X}$ is determined by the three parameters $(\tilde{\alpha}, \tilde{c}, \tilde{\mu})$ or $(\tilde{\alpha}, \tilde{\rho}, \tilde{\mu})$. Since

$$\log E[e^{ivX(t)}] = t\{ivc + \alpha(E[e^{-ivZ_j}] - 1)\}$$

$$= t\left\{ivc + \alpha\left(1 - iv\zeta_1 - \frac{v^2}{2}\zeta_2 + i\frac{v^3}{6}\zeta_3 + o(v^3) - 1\right)\right\}$$

$$= t\left\{iv(c - \alpha\zeta_1) - \frac{v^2}{2}\alpha\zeta_2 + i\frac{v^3}{6}\alpha\zeta_3 + o(v^3)\right\}$$

we have (Cramér 1945, p. 186)

$$E[X(t)] = (c - \alpha\zeta_1)t = \rho\alpha\zeta_1 t \quad \text{(as we already know)},$$
$$E[X^2(t)] = \alpha\zeta_2 t + (\rho\alpha\zeta_1 t)^2,$$
$$E[X^3(t)] = -\alpha\zeta_3 t + 3(\rho\alpha\zeta_1 t)(\alpha\zeta_2 t) + (\rho\alpha\zeta_1 t)^3.$$

Thus the parameters $(\tilde{\alpha}, \tilde{\rho}, \tilde{\mu})$ must satisfy

$$\rho\alpha\zeta_1 = \tilde{\rho}\tilde{\alpha}\tilde{\mu}, \qquad \alpha\zeta_2 = 2\tilde{\alpha}\tilde{\mu}^2, \qquad \alpha\zeta_3 = 6\tilde{\alpha}\tilde{\mu}^3$$

and we get

$$\tilde{\mu} = \frac{\zeta_3}{3\zeta_2}, \qquad \tilde{\rho} = \frac{2\zeta_1\zeta_3}{3\zeta_2^2}\rho \qquad \left(\text{and } \tilde{\alpha} = \frac{9\zeta_2^3}{2\zeta_3^2}\alpha\right).$$

Thus we are led to the approximation

$$\Psi(u) \approx \Psi_{DV}(u) = \frac{1}{1+\tilde{\rho}} e^{-\frac{u\tilde{\rho}}{\tilde{\mu}(1+\tilde{\rho})}}. \tag{31}$$

By construction it follows that $\Psi_{DV}(u) = \Psi(u)$ in the case of exponentially distributed claims.

When Assumption 4 holds it follows from (**III**) and (**IV**) that the Lundberg exponent R is of fundamental importance for the behavior of the ruin probability. It is natural, see (27), to consider

$$R_D = \frac{2\rho\mu}{\mu^2 + \sigma^2} = \frac{2\zeta_1}{\zeta_2}\rho \tag{32}$$

as the "diffusion approximation" of R. With this notation Remark 17 says that $R_D \geq R$. In the same way it is natural, see (31), to put

$$R_{DV} = \frac{\tilde{\rho}}{\tilde{\mu}(1+\tilde{\rho})} = \frac{2\zeta_1}{\zeta_2 + \frac{2\zeta_1\zeta_3}{3\zeta_2}\rho}\,\rho. \qquad (33)$$

The Beekman-Bowers approximation (30) is not exactly exponential, but it seems natural to put R_{BB} equal to the scale parameter in G. This means that

$$R_{BB} = \frac{\mu_B}{\sigma_B^2} = \frac{2\zeta_1}{\zeta_2 + (\frac{4\zeta_1\zeta_3}{3\zeta_2} - 1)\rho}\,\rho. \qquad (34)$$

Numerical comparison of the different approximations are found in, for example, Grandell and Segerdahl (1971), Grandell (1977), Grandell (1978), and De Vylder (1978). In the illustrations to be given, we always put $\alpha = 1$ and $\mu = 1$.

EXAMPLE 19. Γ-DISTRIBUTED CLAIMS. As our first example we consider the case with $\rho = 10\%$ and where the claims are Γ-distributed with $\sigma^2 = 100$. Then we have $\zeta_1 = 1$, $\zeta_2 = 101$, and $\zeta_3 = 20301$. Further,

$$\begin{aligned} R &= 0.0017450, \\ R_D &= 0.0019802, \\ R_{BB} &= 0.0016992, \\ R_{DV} &= 0.0017483. \end{aligned}$$

In Table 1 we give $\Psi(u)$ and the relative errors for the different approximations.

TABLE 1. Γ-distributed claims. $\rho = 10\%$ and $\sigma^2 = 100$.

u	$\Psi(u)$	$\mathcal{E}_D$	$\mathcal{E}_{BB}$	$\mathcal{E}_{DV}$
300	0.52114	5.9%	− 0.1%	0.3%
600	0.30867	− 1.3%	− 0.8%	0.2%
900	0.18287	− 8.0%	− 0.9%	0.1%
1200	0.10834	− 14.3%	− 0.7%	− 0.0%
1500	0.06418	− 20.1%	− 0.2%	− 0.1%
1800	0.03803	− 25.5%	0.3%	− 0.2%
2100	0.02253	− 30.6%	1.0%	− 0.3%
2400	0.01335	− 35.4%	1.8%	− 0.4%
2700	0.00791	− 39.8%	2.7%	− 0.5%
3000	0.00468	− 43.8%	3.6%	− 0.5%

We have not given the relative errors for the Cramér-Lundberg approximation, since – except for $\Psi_{CL}(300) = 0.52100$ and $\Psi_{CL}(600) = 0.30866$ – they coincide with the exact value within the given accuracy.

Grandell and Segerdahl (1971) also considered the case $\sigma^2 = 1000$ and u equal to ten times the values given in Table 1. We shall not give these figures, since for large values of σ^2 both the exact and the approximative ruin probabilities essentially only depends on σ^2/u, and thus such a table should show the same pattern as Table 1.
□

The most striking impression of Table 1 is certainly the extremely good accuracy of the simple De Vylder approximation.

It is a natural question if this is a coincidence (e.g., since R_{DV} "happens" to be very close to R) or if the De Vylder approximation really is so good. We shall therefore consider another claim distribution, where Assumption 4 holds.

EXAMPLE 20. MIXED EXPONENTIALLY DISTRIBUTED CLAIMS. The claim distribution

$$F(z) = 1 - 0.0039793e^{-0.014631z} -$$
$$0.1078392e^{-0.190206z} - 0.8881815e^{-5.514588z} \quad (35)$$

for $z \geq 0$ has been considered by Wikstad (1971). This distribution is an attempt to interpret a distribution (Swedish non-industry fire insurance) considerd by Cramér (1955, pp. 43 - 45). For this distribution we have $\zeta_1 = 1$, $\zeta_2 = 43.1982$, and $\zeta_3 = 7717.23$. We consider those values of u and ρ treated by Wikstad (1971). In Table 2 the Rs and their approximations are given.

TABLE 2. Mixed exponentially distributed claims.

ρ	R	R_D	R_{DV}
5%	0.0020	0.0023	0.0020
10%	0.0036	0.0046	0.0036
15%	0.0049	0.0069	0.0049
20%	0.0059	0.0093	0.0060
25%	0.0067	0.0116	0.0069
30%	0.0074	0.0139	0.0076
100%	0.0114	0.0162	0.0082

From Table 2 it is seen that R_{DV} almost perfectly approximates R for $\rho \leq 30\%$.

The approximations Ψ_D and Ψ_{DV} are simply calculated. In order to compare with the Cramér-Lundberg approximation Ψ_{CL} we put

$$F(z) = \sum_{k=1}^{3}(1 - e^{-z/\mu_k})p_k.$$

Then we have

$$h(r) = \sum_{k=1}^{3} \frac{p_k}{1 - \mu_k r} - 1$$

and

$$h'(r) = \sum_{k=1}^{3} \frac{p_k \mu_k}{(1 - \mu_k r)^2}.$$

Thus, see (III), the Cramér-Lundberg approximation is given by

$$\Psi_{CL}(u) = \frac{\rho}{h'(R) - 1 - \rho} e^{-Ru}.$$

TABLE 3. Mixed exponentially distributed claims.

u	ρ	$\Psi(u)$	$\mathcal{E}_{CL}$	$\mathcal{E}_D$	$\mathcal{E}_{DV}$
10	5%	0.8897	– 3.6%	9.8%	– 3.2%
10	10%	0.7993	– 6.7%	19.4%	– 5.4%
10	15%	0.7242	– 9.2%	28.8%	– 7.0%
10	20%	0.6611	– 11.4%	37.9%	– 8.1%
10	25%	0.6073	– 13.2%	46.7%	– 9.0%
10	30%	0.5610	– 14.8%	55.1%	– 9.6%
10	100%	0.2634	– 23.3%	139.0%	– 10.7%
100	5%	0.7144		11.1%	0.4%
100	10%	0.5393		16.7%	1.1%
100	15%	0.4247		17.6%	1.9%
100	20%	0.3455		14.7%	2.7%
100	25%	0.2886		8.9%	3.4%
100	30%	0.2461		1.3%	4.0%
100	100%	0.0724		– 86.5%	7.2%
1000	5%	0.1149		– 14.0%	0.0%
1000	10%	0.0210		– 53.5%	– 0.9%

In the calculation of the $\Psi_{CL}(u)$-values, we have used more accurate R-values, than those given in Table 2. For $u = 100$ and 1000, $\Psi(u)$ and $\Psi_{CL}(u)$ differ at most by one unit in the last decimal given in Table 3. □

Also in this case the De Vylder approximation must be said to work very well, at least for u not too small and ρ not too large.

As our last example we consider lognormally distributed claims. This is a case where Assumption 4 does not hold.

EXAMPLE 21. LOGNORMALLY DISTRIBUTED CLAIMS. A random variable Z is lognormally distributed if $\log Z$ is normally distributed. Put

$$\mu_L = E[\log Z] \quad \text{and} \quad \sigma_L^2 = \text{Var}[\log Z].$$

We have

$$\zeta_k = e^{k\mu_L + \frac{1}{2}k^2\sigma_L^2}$$

and thus the normalization $\zeta_1 = 1$ implies $\mu_L = -\frac{1}{2}\sigma_L^2$. Thorin and Wikstad (1977) considered $\sigma_L = 1.8$. That choice of σ_L was taken from Benckert and Jung (1974). They found that value in their investigation of the Swedish experience of fire insurance of stone dwellings reported from 1958 - 1969. In this case we have $\zeta_1 = 1$, $\zeta_2 = 25.53372$, and $\zeta_3 = 16647.24$.

TABLE 4. Lognormally distributed claims.

u	ρ	$\Psi(u)$	$\mathcal{E}_D$	$\mathcal{E}_{DV}$
100	5%	0.55074	22.7%	− 20.6%
100	10%	0.34395	32.8%	− 19.5%
100	15%	0.23573	31.0%	− 14.2%
100	20%	0.17309	20.6%	− 8.1%
100	25%	0.13384	5.4%	− 2.1%
100	30%	0.10765	− 11.4%	3.5%
100	100%	0.02535	− 98.4%	41.7%
1000	5%	0.04199	− 52.6%	55.1%
1000	10%	0.01099	− 96.4%	85.5%
1000	15%	0.00574	− 99.9%	79.7%
1000	20%	0.00384	− 100.0%	68.7%
1000	25%	0.00288	− 100.0%	59.2%
1000	30%	0.00230	− 100.0%	51.8%

In this case the approximations work bad. Maybe this is not too surprising, since in this case (Thorin and Wikstad 1977, p. 243)

$$\Psi(u) \sim \frac{\sigma_L^3}{\rho\sqrt{2\pi}\log(u\sqrt{\zeta_2})\log(u/\sqrt{\zeta_2})} e^{-\frac{1}{2\sigma_L^2}(\log(u/\sqrt{\zeta_2}))^2}$$

while the approximations are exponentially decreasing. A good survey of the "non-exponential" case is given by Embrechts and Veraverbeke (1982).

The fact that De Vylder's approximation works well for $u = 100$ and $\rho = 25\%$ and 30% is probably just a coincidence.
□

One – but important – reason to look for simple approximations is that they are useful for finding "rule of thumbs" to decide if a risk process is on "the safe side." The intensity α and the distribution F (or the moments $\zeta_1, \zeta_2,$ and ζ_3) may be regarded as parameters characterizing the portfolio while ρ and u may be regarded as decision variables. If a ruin probability p is acceped it is natural to choose ρ and u such that $\Psi(u) = p$. In practice Ψ has to be replaced by some approximation. Using the diffusion approximation we get

$$u = -\frac{\log p}{R_D}$$

and using De Vylder's approximation

$$u = -\frac{\log p + \log(1 + \tilde{\rho})}{R_{DV}}.$$

Such simple "rule of thumbs" are not produced by Beekman-Bowers' approximation, since it requires computations of the Γ-distribution function $G(u)$. For some of the u-values used in our examples the values of $G(u)$ were numerically uncertain when the PC statistical package *Statgraphics* were used. Although this numerical uncertainty might be easily overcome it is obvious that De Vylder's approximation is more attractive for practical use than Beekman-Bowers' approximation. The good accuracy indicated in our examples of De Vylder's approximation, at least when Assumption 4 holds, is also shown in the two further comparisons given by De Vylder (1978). Our impression is that the title of his paper has relevance.

If we compare the diffusion approximation with De Vylder's approximation, the diffusion approximation may look like "the ugly duckling." Ugly, of course, since the comparisons indicate very bad accuracy. The diffusion approximation may be looked upon as if the risk process is replaced by a Wiener process with drift (where the first two moments coincide). The De Vylder approximation, on the other hand, is obtained by replacement of the risk process with a risk process with exponentially distributed claims (such that the first three moments coincide). Very intuitively the "De Vylder replacement" ought to be a smaller change than the "diffusion replacement," since a further moment is involved and since a risk process ought to be "closer" to another risk process than to a Wiener process. Therefore it is not surprising that the De Vylder approximation works better than the diffusion approximation.

In the fairy tale, the duckling is a swan. From a glance at the tables, one is hardly willing to look upon the diffusion approximation as a swan. It does, however, have some nice properties.

1. It is based on a limit theorem. From a practical point of view this is a poor consolation.

2. The underlying limit theorem $\bar{S}_n \Rightarrow W$ holds for much more general point processes than the Poisson process. In general, we do, however, not know when this implies $P\{\inf_{t \geq 0} Y_n(t) < -y\} \to P\{\inf_{t \geq 0} Y(t) < -y\}$. If we consider "ruin within finite time" this is no problem. For a discussion about these questions we refer to Grandell (1977) and the discussion in Section A.4.

3. Asmussen (1984) applied the diffusion approximation to the so-called conjugate process and achieved much better accuracy. The construction of that process requires full knowledge of F, and thus the simplicity of the ordinary diffusion approximation is lost.

1.3 Inference for the risk process

We shall now consider the situation, where we want to apply risk theory to a specific portfolio. In general neither α nor $F(z)$ are known. One method, often used, is to try to fit some model to actual data, and to then make the desired calculation in the fitted model. This is, essentially, how the specific mixed exponential and lognormal distributions, used in the comparison in Section 1.2, were obtained.

This way of fitting a model is not quite satisfying, since the results give no indication of how certain the model is. Naturally these kind of objections can always be raised, since almost all applications of theoretical results to specific situations are based on combinations of assumptions and observations. If we, for example, want to use De Vylder's approximation we have to assume that the claims follow a Poisson process and to estimate α, ζ_1, ζ_2, and ζ_3. Since those parameters are simply estimated and since we are using an approximation we do not believe that too much objection can be raised towards the above approach. If we, on the other hand, want to use the Cramér-Lundberg approximation or the Lundberg inequality, the situation is slightly different since we are probably interested in the ruin probability for large values of u. The problem is that one only has observations in the "probable" area of $F(z)$ while the ruin probabilty depends highly on the "tail" of $F(z)$.

We shall consider a different approach, discussed by Grandell (1979). Our estimates are based on an observation of the risk process $X(t)$ for $t \in [0, T]$. Roughly speaking we shall just replace $F(z)$ by the empirical distribution function and α with its natural estimate. Of course, we too cannot estimate the "tail" of $F(z)$, but the problem is at least not hidden.

DEFINITION 22. We say that we are in the *regular* case if

(i)   The claims follow a Poisson process;

(ii) Assumption 4 holds;

(iii) $c > \alpha\mu$, i.e., we have positive safety loading;

(iv) $h(2R) < \infty$.

Consider the regular case and recall that R is the positive solution of $h(r) = cr/\alpha$. It is practical to introduce the function $g(r)$, defined by

$$g(r) \stackrel{\text{def}}{=} h(r) - \frac{cr}{\alpha} = \int_0^\infty e^{rz}\, dF(z) - 1 - \frac{cr}{\alpha},$$

since then R is the positive solution of $g(r) = 0$. Consider the risk process $X(t)$ for $t \in [0,T]$ and define the random process

$$G_T(r) = \frac{1}{N(T)} \sum_{k=1}^{N(T)} e^{rZ_k} - 1 - cr \left(\frac{N(T)}{T}\right)^{-1} \qquad \text{if } N(T) > 0.$$

Replacing $X(t)$ by an observation $x(t)$ we can form the corresponding function $g_T(r)$. If $x(T) > 0$ and if at least one claim has occurred, this function has the same properties as $g(r)$ and a natural estimate of R is given by the positive solution R^* of $g_T(r) = 0$. In order to study the properties of R^* we define the random variable R_T as the positive solution of $G_T(r) = 0$ when such a solution exists.

REMARK 23. There is always a positive probability that

$$c \leq \frac{N(T)}{T} \frac{1}{N(T)} \sum_{k=1}^{N(T)} Z_k,$$

i.e., that $X(T) \leq 0$, or that $N(T) = 0$. In those cases we put $R_T = 0$ and $R_T = +\infty$, respectively.

We make the corresponding convention for R^*, although it is hardly necessary. In practice no one will try to make any estimation before claims have occurred. Further, if $x(T) \leq 0$ the company has probably more acute problems than statistical estimation, or wants to consider the ruin probability for a higher gross risk premium.
□

Our basic result is the following theorem.

THEOREM 24. *In the regular case*

$$\sqrt{T}(R_T - R) \stackrel{d}{\to} Y \qquad \text{as } T \to \infty,$$

where Y is a normally distributed random variable with $E[Y] = 0$ and

$$\sigma_Y^2 \stackrel{\text{def}}{=} \mathrm{Var}[Y] = \frac{g(2R)}{\alpha(g'(R))^2} = \frac{h(2R) - 2cR/\alpha}{\alpha(h'(R) - c/\alpha)^2}.$$

Before proving the theorem we shall give a lemma.

LEMMA 25. *In the regular case*
$$\frac{G_T(R)}{R_T - R} \to -g'(R) \quad P\text{-a.s.} \quad \text{as } T \to \infty.$$

PROOF OF LEMMA 25: The proof is similar to the proof of asymptotic normality of maximum likelihood estimates given by Cramér (1945, pp. 500 - 503).

All statements about random quantities are meant to hold P-a.s.

Since $N(T)/T \to \alpha$ as $t \to \infty$ and since $E[e^{rZ_k}] < \infty$ for $r \leq 2R$ it follows for $r < 2R$ that $G_T(r) \to g(r)$ and $G_T'(r) \to g'(r)$ as $T \to \infty$.

In the regular case $g'(R) > 0$. Choose $\epsilon \in (0, R)$ such that $g'(R - \epsilon) > 0$. For T (depending on the realization of $X(t)$ and on ϵ) large enough we have $G_T(R - \epsilon) < 0$, $G_T(R + \epsilon) > 0$, and $G_T'(R - \epsilon) > 0$. Thus $|R_T - R| < \epsilon$.

Now $G_T(R) = G_T(R) - G_T(R_T) = -(R_T - R)G_T'(R + \theta_T(R_T - R))$ for some $\theta_T \in (0, 1)$. Thus
$$\frac{G_T(R)}{R_T - R} = -G_T'(R + \theta_T(R_T - R)),$$
provided $G_T(R) \neq 0$. If $G_T(R) = 0$ we have $R_T = R$ and then we just define the ratio as $-G_T'(R)$. Since
$$G_T''(r) = \frac{1}{N(T)} \sum_{k=1}^{N(T)} Z_k^2 e^{rZ_k} > 0$$
it follows that $G_T'(r)$ is increasing in r and we have
$$|G_T'(R + \theta_T(R_T - R)) - g'(R)| \leq |G_T'(R - \epsilon) - g'(R)| + |G_T'(R + \epsilon) - g'(R)|$$
$$\to |g'(R - \epsilon) - g'(R)| + |g'(R + \epsilon) - g'(R)| \quad \text{as } T \to \infty$$
which can be made arbitrarily small by choosing ϵ small enough. ∎

PROOF OF THEOREM 24: We have
$$G_T(R) = G_T(R) - g(R) = \frac{1}{N(T)} \sum_{k=1}^{N(T)} e^{RZ_k} - 1 - h(R) - cR\left(\frac{T}{N(T)} - \frac{1}{\alpha}\right).$$

Let S_k denote the epoch of the kth claim and put $S_0 = 0$. The variables $S_1 - S_0, S_2 - S_1, S_3 - S_2,\ldots$ are independent and exponentially distributed with mean $1/\alpha$. We have
$$T = S_{N(T)} + (T - S_{N(T)}) = \sum_{k=1}^{N(T)} (S_k - S_{k-1}) + (T - S_{N(T)})$$
and thus
$$\frac{T}{N(T)} - \frac{1}{\alpha} = \frac{1}{N(T)} \sum_{k=1}^{N(T)} \left((S_k - S_{k-1}) - \frac{1}{\alpha}\right) + \frac{T - S_{N(T)}}{N(T)}.$$

Thus
$$\sqrt{T}\, G_T(R) = \sqrt{\frac{T}{N(T)}} \cdot \frac{1}{\sqrt{N(T)}} \cdot$$
$$\sum_{k=1}^{N(T)} \left[\left\{(e^{rZ_k} - 1 - h(R)) - cR\left((S_k - S_{k-1}) - \frac{1}{\alpha}\right)\right\} - cR\,\frac{T - S_{N(T)}}{N(T)}\right].$$

Now $N(T)/T \to \alpha$ and $T - S_{N(T)} \xrightarrow{d}$ an exponentially distributed random variable. The random variables
$$\{[e^{rZ_k} - 1 - h(R)] - cR[(S_k - S_{k-1}) - 1/\alpha]\}, \quad k = 1, 2, \ldots$$
are independent with means zero and variances
$$h(2R) + 1 - (h(R)+1)^2 + \left(\frac{cR}{\alpha}\right)^2 = h(2R) - \frac{2cR}{\alpha} = g(2R).$$

From all this and the classical generalization of the central limit theorem to sums of a random number of random variables, see, e.g., Rényi (1960, p. 98), it follows that
$$\sqrt{T}\, G_T(R) \xrightarrow{d} \sqrt{\frac{1}{\alpha}}\,[\sqrt{g(2R)} \cdot W + 0],$$
where W is a normally distributed random variable with mean zero and variance one.

Since $0 < g'(R) < \infty$ it follows from Lemma 25 that
$$\sqrt{T}\,(R_T - R) = \sqrt{T}\, G_T(R) \cdot \frac{R_T - R}{G_T(R)} \xrightarrow{d} \sqrt{\frac{1}{\alpha}}\,\sqrt{g(2R)} \cdot \frac{1}{g'(R)} \cdot W$$
which equals Y in distribution. ∎

Theorem 24 can be used to form confidence intervals for R. In practice σ_Y is unknown and we have to replace it by its natural estimate
$$\sigma_Y^* = \frac{\sqrt{g_T(2R^*)}}{\sqrt{\alpha^*}\, g_T'(R^*)},$$
where $\alpha^* = n(T)/T$.

A one-sided approximate 95% confidence interval for R is thus given by
$$\left(R^* - \frac{1.6\sigma_Y^*}{\sqrt{T}},\, \infty\right).$$

This interval leads us to the following empirical Lundberg inequality
$$\Psi(u) \leq e^{-(R^* - 1.6\sigma_Y^*/\sqrt{T})u} \tag{36}$$
which holds for all u in approximately 95% of all investigations.

In many situations we may be more interested in an estimate of the ruin probability than in an inequality. When our interest is in large values of

u it is natural to use the Cramér-Lundberg approximation (**III**), which in the notation used here is given by

$$\Psi_{CL}(u) = Ce^{-Ru}, \quad \text{where } C = \frac{c - \alpha\mu}{\alpha g'(R)}$$

for such an estimate. A natural estimate of μ is $\mu^* = (cT - x(T))/n(T)$ and thus a natural estimate of C is $C^* = x(T)/[n(T)g'_T(R^*)]$. Define the estimate $\Psi^*(u)$ and the random variable $\Psi_T(u)$ by

$$\Psi^*(u) \stackrel{\text{def}}{=} \frac{x(T)}{n(T)g'_T(R^*)} e^{-R^*u} = C^* e^{-R^*u}$$

and

$$\Psi_T(u) \stackrel{\text{def}}{=} \frac{X(T)}{N(T)G'_T(R_T)} e^{-R_T u} = C_T e^{-R_T u}.$$

Consider the "relative error" $\mathcal{E}_T(u)$ defined by

$$\mathcal{E}_T(u) = \frac{\Psi_T(u) - \Psi(u)}{\Psi(u)} = \frac{\Psi_T(u)}{\Psi_{CL}(u)} \cdot \frac{\Psi_{CL}(u)}{\Psi(u)} - 1$$

and note that $\mathcal{E}_T(u)$ is a random variable "containing" both the error in the Cramér-Lundberg approximation and the "random" error.

Thus we have

$$\log(\mathcal{E}_T(u) + 1) = \log \frac{C_T}{C} - u(R_T - R) + \log \frac{\Psi_{CL}(u)}{\Psi(u)}.$$

Since $\log(C_T/C) \to 0$ as $T \to \infty$ P-a.s. and $\log[\Psi_{CL}(u)/\Psi(u)] \to 0$ as $u \to \infty$ it is natural to let $T \to \infty$ together with u in such a way that $u/\sqrt{T} \to \tilde{u} \in (0, \infty)$. From Theorem 24 we then get

$$\log(\mathcal{E}_T(\tilde{u}\sqrt{T}) + 1) \stackrel{d}{\to} \tilde{u}\, Y \quad \text{as } T \to \infty. \tag{37}$$

In the same way as (36) follows from Theorem 24 it follows from (37) that

$$\left(C^* e^{-(R^*+2\sigma^*_Y/\sqrt{T})u},\ C^* e^{-(R^*-2\sigma^*_Y/\sqrt{T})u}\right) \tag{38}$$

is an approximate 95% confidence interval for $\Psi(u)$ when u and $\sqrt{T}$ are of the same large order.

As we have mentioned, the ruin probabilty highly depends on the "tail" of $F(z)$ for large values of u. The larger T is the more information we get about the "tail," is formalized by the requirement that u and $\sqrt{T}$ must be of the same order.

Because of the construction of (38) we may consider all u larger than some u_0 simultaneously without changing the level, provided that u_0 and $\sqrt{T}$ are of the same large order. To realize this we consider the random variable

$$\sup_{u \geq u_0} \left| \frac{\sqrt{T}}{u} \log(\mathcal{E}_T(u) + 1) \right|$$

$$= \sup_{u \geq u_0} \left| \frac{\sqrt{T}}{u} \log \frac{C_T}{C} - \sqrt{T}(R_T - R) + \frac{\sqrt{T}}{u} \log \frac{\Psi_{CL}(u)}{\Psi(u)} \right|$$

$$\leq \sup_{u \geq u_0} \left| \frac{\sqrt{T}}{u} \log \frac{C_T}{C} \right| + |\sqrt{T}(R_T - R)| + \sup_{u \geq u_0} \left| \frac{\sqrt{T}}{u} \log \frac{\Psi_{CL}(u)}{\Psi(u)} \right| \xrightarrow{d} |Y|$$

as $T \to \infty$, $u_0 \to \infty$, and $u/\sqrt{T} \to \tilde{u}_0 \in (0, \infty)$.
Thus

$$0.95 \approx P\left\{ \left| \frac{\sqrt{T}}{u} \log(\mathcal{E}_T(u) + 1) \right| \leq 2\sigma_Y \quad \text{for all } u \geq u_0 \right\}$$

$$= P\{|\log \Psi(u) - \log(C_T e^{-R_T u})| \leq 2u\sigma_Y/\sqrt{T} \quad \text{for all } u \geq u_0\}$$

and it follows that all $u \geq u_0$ may be considered simultaneously.

EXAMPLE 26. Consider the case when Z_k is exponentially distributed. Then $h(R) < \infty$ for $c < 2\alpha\mu$ or $\rho < 100\%$. In that case we have

$$\sigma_Y^2 = \frac{2}{\alpha\mu^2(1+\rho)^2(1-\rho)}.$$

It is natural to ask what happens if $\rho > 100\%$. In this case Theorem 24 does not hold any more. Lemma 25 does, however, still hold. From the lemma and from the theory of stable distributions, see Feller (1971, pp. 570, 577, 581), it follows that

$$T^{1/(1+\rho)}(R_T - R) \xrightarrow{d} Y_\rho \quad \text{as } T \to \infty,$$

where Y_ρ has a stable distribution with exponent $(1 + \rho)/\rho$. The characteristic function for Y_ρ can be calculated, but it is so complicated that the result is of no practical interest.
□

REMARK 27. The fact that R^* is the positive solution $g_T(r) = 0$, where

$$g_T(r) = \frac{1}{n(T)} \sum_{k=1}^{n(T)} e^{rz_k} - 1 - cr\left(\frac{n(T)}{T}\right)^{-1},$$

may be regarded as a practical drawback of this method of estimation since the numerical problems of computing R^* may be expected to be considerable.

Rosenlund (1989) has applied this method of estimating R on real claim statistics, consisting of 182,342 claims, at the Swedish insurance company "Länsförsäkringsbolagen." He solved the equation $g_T(r) = 0$ with the secant method, i.e., the Newton-Raphson method with the derivative replaced by a difference ratio. With 9 computations of $g_T(r)$ the total CPU time on an IBM 3090 was only 14.6 sec. Thus the numerical problems are almost negligible.
□

Let us consider the simplified, but less natural, situation where α is known and where the first n claims are observed. Then we consider the random process

$$G_n(r) = \frac{1}{n}\sum_{k=1}^{n} e^{rZ_k} - 1 - \frac{cr}{\alpha}$$

and define R_n as the positive solution of $G_n(r) = 0$. Define, for future purpose, the random variable $H_k(r)$ by

$$H_k(r) = e^{rZ_k} - 1 - \frac{cr}{\alpha}$$

and note that

$$E[H_k(r)] = g(r) \quad \text{and} \quad \sigma_H^2 \stackrel{\text{def}}{=} \text{Var}[H_k(R)] = g(2R) - \left(\frac{cR}{\alpha}\right)^2,$$

By obvious modifications of the proof of Theorem 24 it follows that

$$\sqrt{n}(R_n - R) \stackrel{d}{\to} \frac{\sigma_H}{g'(R)} \cdot W \quad \text{as } n \to \infty. \tag{39}$$

Herkenrath (1986) considers estimation of R as a stochastic approximation problem and proposes a modified Robbins-Monro procedure for its solution. The idea behind this approach can roughly be described in the following way. Let $\hat{R}_0$ be a starting value, or an "initial estimate" of R when no claim has occurred. Let $\hat{R}_k$ be the estimate based on the first k claims. When the $(k+1)$th claim Z_{k+1} occurs, we want to form $\hat{R}_{k+1}$ recursively, i.e., $\hat{R}_{k+1}$ shall depend only on $\hat{R}_k$ and Z_{k+1}.

Consider now the function $g(r)$. We know that

$$g(0) = 0, \quad g(r) < 0 \text{ for } 0 < r < R, \quad g(R) = 0 \text{ and}$$
$$0 < g(r) < \infty \text{ for } R < r < r_\infty.$$

Assume that we can find, or believe in, an interval $[R_{\min}, R_{\max}]$ such that

$$0 < R_{\min} < R < R_{\max} < r_\infty,$$

to which the estimates are restricted. It then seems natural, forgetting for the moment about the restriction to $[R_{\min}, R_{\max}]$, to put

$$\hat{R}_{k+1} = \hat{R}_k - a_k H_{k+1}(\hat{R}_k).$$

Since $g(r) < (>) 0$ for $r < (>) R$ it is natural to require that $a_k > 0$. Further, the "additional" information in Z_{k+1} compared to $\hat{R}_k$ decreases and thus it is natural to require that $a_k \searrow 0$ as $k \to \infty$. One such choice, which works well, is to choose $a_k = a/k$ where $a > 0$. Finally we shall restrict the estimates to $[R_{\min}, R_{\max}]$ and we are led to

$$\hat{R}_{k+1} = \begin{cases} R_{\min} & \text{if } \hat{R}_k - \frac{a}{k}H_{k+1}(\hat{R}_k) < R_{\min} \\ \hat{R}_{k+1} = \hat{R}_k - \frac{a}{k}H_{k+1}(\hat{R}_k) & \text{if } R_{\min} \leq \hat{R}_k - \frac{a}{k}H_{k+1}(\hat{R}_k) \leq R_{\max} \\ R_{\max} & \text{if } R_{\max} < \hat{R}_k - \frac{a}{k}H_{k+1}(\hat{R}_k) \end{cases}.$$

Under some additional conditions on $g(r)$ we have (Sacks 1958, p. 383)

$$\sqrt{n}(\hat{R}_n - R) \xrightarrow{d} \frac{a\sigma_H}{\sqrt{2ag'(R) - 1}} \cdot W \qquad \text{as } n \to \infty \qquad (40)$$

provided that $a > 1/(2g'(R))$. It is easily seen that $a = 1/g'(R)$ is (asymptotically) optimal. In that case, see (39) and (40), R_n and $\hat{R}_n$ have the same asymptotic behavior. Thus the estimate R^* seems preferable compared to $\hat{R}_n$. In our opinion this conclusion is intuitively natural, but it is *not* in agreement with the conclusions drawn by Herkenrath (1986). His conclusions are based on a simulation study.

CHAPTER 2

Generalizations of the classical risk model

There are certainly many directions in which the classical risk model needs generalization in order to become a reasonably realistic description of the actual behavior of a risk movement. We shall, almost solely, consider generalizations where the occurrence of the claims is described by point processes other than the Poisson process. This restriction is more a reflection of our personal interest than an ambition to cover the most important aspects of risk theory.

There are, at least, two very different reasons for using other models for the claim occurrence than the Poisson process. *First* the Poisson process is stationary, which – among other things – implies that the number of policyholders involved in the portfolio cannot increase (or decrease). Few insurance managers would accept a model where the possibility of an increase of the business is not taken into account. We shall refer to this case as *size fluctuation*. *Second* there may be fluctuation in the underlying risk. Typical examples are automobile insurance and fire insurance. We shall refer to this as *risk fluctuation*.

2.1 Models allowing for size fluctuation

The simplest way to take size fluctuation into account is to let N be a non-homogeneous Poisson process. Let $A(t)$ be a continuous non-decreasing function with $A(0) = 0$ and $A(t) < \infty$ for each $t < \infty$.

DEFINITION 1. *A point process N is called a (non-homogeneous) Poisson process with intensity measure A if*

(i) $N(t)$ has independent increments;

(ii) $N(t) - N(s)$ is Poisson distributed with mean $A(t) - A(s)$.

REMARK 2. The function $A(t)$ can be looked upon as the distribution function corresponding to the measure A. The continuity of $A(\cdot)$ guarantees that N is *simple*, i.e., that $N(\cdot)$ increases exactly one unit at its epochs of increase. □

Define the inverse A^{-1} of A by

$$A^{-1}(t) = \sup(s \mid A(s) \leq t). \tag{1}$$

A^{-1} is always right-continuous. Since $A(\cdot)$ is continuous, A^{-1} is (strictly) increasing and

$$A \circ A^{-1}(t) \stackrel{\text{def}}{=} A(A^{-1}(t)) = t \quad \text{for } t < A(\infty).$$

DEFINITION 3. A Poisson process $\tilde{N}$ with $\alpha = 1$ is called a *standard* Poisson process.

The following obvious results are, due to their importance, given as lemmata.

LEMMA 4. *Let N be a Poisson process with intensity measure A such that $A(\infty) = \infty$. Then the point process $\tilde{N} \stackrel{\text{def}}{=} N \circ A^{-1}$ is a standard Poisson process.*

PROOF: Since A^{-1} is increasing it follows that $\tilde{N}$ has independent increments. Further, $\tilde{N}(t) - \tilde{N}(s) = N(A^{-1}(t)) - N(A^{-1}(s))$ is Poisson distributed with mean $A \circ A^{-1}(t) - A \circ A^{-1}(s) = t - s$. ∎

LEMMA 5. *Let $\tilde{N}$ be a standard Poisson process. Then the point process $N \stackrel{\text{def}}{=} \tilde{N} \circ A$ is a Poisson process with intensity measure A.*

The proof is omitted.

Without much loss of generality we may assume, although it is not at all necessary, that A has the representation

$$A(t) = \int_0^t \alpha(s) \, ds, \tag{2}$$

where $\alpha(\cdot)$ is called the *intensity function*. It is natural to assume that $\alpha(s)$ is proportional to the number of policyholders at time s. When the premium is determined individually for each policyholder it is also natural to assume the gross risk premium to be proportional to the number of policyholders. If the relative safety loading ρ is constant we get $c(t) = (1 + \rho)\mu\alpha(t)$ and the corresponding risk process is given by, see (1.1),

$$X(t) = (1 + \rho)\mu A(t) - \sum_{k=1}^{N(t)} Z_k,$$

where N is a Poisson process with intensity measure A such that $A(\infty) = \infty$.

Consider now the process $\tilde{X}$ defined by

$$\tilde{X}(t) \stackrel{\text{def}}{=} X \circ A^{-1}(t) = (1+\rho)\mu t - \sum_{k=1}^{\tilde{N}(t)} Z_k.$$

Thus $\tilde{X}$ is a classical risk process with $\alpha = 1$. Recall that

$$\Psi(u) = P\{\inf_{t \geq 0} X(t) < -u\}.$$

If $A(\cdot)$ is increasing, or if $\alpha(t) > 0$, A^{-1} is continuous and it is obvious that $\inf_{t \geq 0} X(t) = \inf_{t \geq 0} \tilde{X}(t)$. Here it would only be a minor restriction to assume that $A(\cdot)$ is increasing, but for the further discussion we do not want to make that restriction. Suppose that A^{-1} has a jump at t. In the time interval $(A^{-1}(t-), A^{-1}(t))$ no claims occur, since $N(A^{-1}(t)) - N(A^{-1}(t-))$ is Poisson distributed with mean $A \circ A^{-1}(t) - A \circ A^{-1}(t-) = t - (t-) = 0$, and no premiums are recieved. Thus $\inf_{t \geq 0} X(t) = \inf_{t \geq 0} \tilde{X}(t)$ and the problem of calculating the ruin probability is brought back to the classical situation.

The time scale defined by A^{-1} is generally called *the operational time scale*, see, e.g., Cramér (1955, p. 19).

We have referred to this generalization as "size fluctuations," only because then the gross risk premium rate $c(t) = (1+\rho)\mu\alpha(t)$ is very natural. Obviously it is mathematically irrelevant *why* $\alpha(\cdot)$ fluctuates, as long as those fluctuations are compensated by the premium in the above way. We shall now see that a kind of operational time scale can be defined for a very wide class of point processes. Those processes may very well more naturally correspond to "risk fluctuation" than to "size fluctuation." Before discussing this wide class we shall introduce Cox processes which are very natural as models for "risk fluctuation." In the sequel they will play an important rôle, although we shall here merely use them as an illustration.

DEFINITION 6. A stochastic process $\Lambda = \{\Lambda(t); t \geq 0\}$ with P-a.s. $\Lambda(0) = 0$, $\Lambda(t) < \infty$ for each $t < \infty$ and non-decreasing realizations is called a *random measure*.

DEFINITION 7. A random measure is called *diffuse* if it has P-a.s. continuous realizations.

DEFINITION 8. Let a random measure Λ and a standard Poisson process $\tilde{N}$ be independent of each other. The point process $N = \tilde{N} \circ \Lambda$ is called a *Cox process* (or a "doubly stochastic Poisson process").

REMARK 9. Definition 8 is one of several equivalent definitions. Strictly speaking we only require that N and $\tilde{N} \circ \Lambda$ are equal in distribution. Further, we ought to show that the mapping $(\tilde{N}, \Lambda) \mapsto \tilde{N} \circ \Lambda$ is measurable. For these questions we refer to Grandell (1976, pp. 9 - 16).
□

Intuitively we shall think of a Cox process N as generated in the following way: First a realization A of a random measure is generated and conditioned upon that realization N is a Poisson process with intensity measure A. A Cox process is simple if and only if the underlying random measure is diffuse.

Let Λ be a diffuse random measure, defined on $(\Omega, \mathcal{F}, P)$, such that $\Lambda(\infty) = \infty$ P-a.s. Let N be the corresponding Cox process. Consider the risk process

$$X(t) = (1+\rho)\mu\Lambda(t) - \sum_{k=1}^{N(t)} Z_k.$$

Define now $\tilde{X} \stackrel{\text{def}}{=} X \circ \Lambda^{-1}$ and the σ-algebra $\mathcal{F}_\infty^\Lambda \stackrel{\text{def}}{=} \sigma\{\Lambda(s);\ s \leq t\}$. From Lemma 4 it follows that $\tilde{N}$ relative to $\mathcal{F}_\infty^\Lambda$, i.e., conditioned upon Λ, is a standard Poisson process. Thus, if $\tilde{\Psi}(u)$ is the ruin probability corresponding to a standard Poisson process, we have

$$\Psi(u) = P\{\inf_{t \geq 0} X(t) < -u\} = P\{\inf_{t \geq 0} \tilde{X}(t) < -u\}$$

$$= E[P^{\mathcal{F}_\infty^\Lambda}\{\inf_{t \geq 0} \tilde{X}(t) < -u\}] = E[\tilde{\Psi}(u)] = \tilde{\Psi}(u). \tag{3}$$

This is certainly by no means surprising, since (3) is just a formal way of saying that a trick which works for *every* intensity also works for a randomly picked out intensity.

Assume now that Λ has the representation

$$\Lambda(t) = \int_0^t \lambda(s)\, ds, \tag{4}$$

where $\lambda = \{\lambda(t);\ t \geq 0\}$ is called the *intensity process*. Obviously $\lambda(t) \geq 0$ P-a.s. Often it is more natural to define a Cox process by specifying λ than by specifying Λ. In all cases to be considered $\lambda(\cdot)$ has right-continuous and Riemann integrable realizations. Then, see Grandell (1976, p. 14), the mapping $\lambda \mapsto \Lambda$, defined by (4), is measurable, and thus the corresponding Cox process is well-defined. The gross risk premium rate is

$$c(t) = (1+\rho)\mu\lambda(t).$$

In order to bring the calculation of $\Psi(u)$ back to the classical case we have, at least, the following problems:

(i)   we must be able to observe $\lambda(\cdot)$;

(ii)   we must be able to continuously change the gross risk premium rate.

Here we shall only discuss problem (i). The intensity process $\lambda(\cdot)$ is a part of the mathematical model, and it is difficult to think of any situation where its realizations really can be observed. Already to talk about an observation of λ might confuse the mathematical model and reality. In practice we therefore must rely on some kind of estimation of $\lambda(t)$. Such estimates may be based on $N(s)$ for $s \leq t$ and, eventually, other external information. Thus, if the estimates are based only on N, it seems natural to replace $\lambda(t)$ with

$$\lambda^*(t) = E^{\mathcal{F}_t^N}[\lambda(t)] \tag{5}$$

and, if also some external information is used, with

$$\hat{\lambda}(t) = E^{\mathcal{F}_t}[\lambda(t)]. \tag{6}$$

The notation $\mathcal{F}_t^N$ and $\mathcal{F}_t$ is explained in Definitions 1.9 and 1.10. We always assume that $\mathcal{F}_t^N \subseteq \mathcal{F}_t$, since – for technical reasons – we want $N(t)$ to be $\mathcal{F}_t$-measurable.

We shall now argue very heuristically, and to emphasize this we put a question mark after the numbers of the formulas. In a simple Cox process

$$\lambda(t)\,dt = P^{\mathcal{F}_\infty^\Lambda}\{dN(t) = 1\} = E^{\mathcal{F}_\infty^\Lambda}[dN(t)]$$

and therefore we can define $\hat{\lambda}(t)$ by

$$\hat{\lambda}(t)\,dt = E^{\mathcal{F}_t}[dN(t)]. \tag{7?}$$

This definition ought to be meaningful for rather general simple point processes, and we do not assume that N is a Cox process anymore. Completely forgetting about "measurability" we can form

$$\hat{\Lambda}(t) = \int_0^t \hat{\lambda}(s)\,ds. \tag{8?}$$

Let us now consider

$$\tilde{N} = N \circ \hat{\Lambda}^{-1}. \tag{9?}$$

Put

$$\mathcal{G}_t = \mathcal{F}_{\hat{\Lambda}^{-1}(t)} \tag{10?}$$

which means that $\mathcal{G}_t$ represents the history up to time $\hat{\Lambda}^{-1}(t)$. Then we have

$$E^{\mathcal{G}_t}[d\tilde{N}(t)] = E^{\mathcal{G}_t}[\tilde{N}(t + dt) - \tilde{N}(t)]$$
$$= E^{\mathcal{F}_{\hat{\Lambda}^{-1}(t)}}[N(\hat{\Lambda}^{-1}(t + dt)) - N(\hat{\Lambda}^{-1}(t))]$$
$$= E^{\mathcal{F}_{\hat{\Lambda}^{-1}(t)}}[N(\hat{\Lambda}^{-1}(t) + dt/\hat{\lambda}(t)) - N(\hat{\Lambda}^{-1}(t))] = \hat{\lambda}(t) \cdot \frac{dt}{\hat{\lambda}(t)} = dt \tag{11?}$$

which implies that

$$\tilde{N} \text{ is a standard Poisson process.} \tag{12?}$$

This highly heuristic reasoning is developed in more detail by Gerber (1979, pp. 25 - 31 and 142 - 143).

The natural way to achieve stringent versions of (8?) and (12?) is to rely on the "martingale approach" to point processes which goes back to

Brémaud (1972). Good references, well suited to our demand, are Brémaud (1981) and Liptser and Shiryayev (1978). Although we shall only state some basic results, we shall – especially concerning Cox processes – take up some properties not needed in this chapter.

POINT PROCESSES AND MARTINGALES

Let a point process N and a (right continuous) filtration **F** be given. We assume P-a.s. that N is simple and that $N(t) < \infty$ for each $t < \infty$. Let $S_1, S_2, \ldots$ denote its jump times. As above we require that N is adapted to **F**, i.e., that $\mathcal{F}_t^N \subseteq \mathcal{F}_t$ for all $t \geq 0$.

In order to get some feeling for how the filtration comes into the theory we redefine the Poisson process. Recall that A is a continuous non-decreasing function with $A(0) = 0$ and $A(t) < \infty$ for each $t < \infty$.

DEFINITION 10. A point process N is called an **F**-Poisson process with intensity measure A if

(i) $N(t) - N(s)$ is independent of $\mathcal{F}_s$;

(ii) $N(t) - N(s)$ is Poisson distributed with mean $A(t) - A(s)$.

Obviously an $\mathbf{F}^N$-Poisson process is a Poisson process according to Definition 1. If, however, $\mathcal{F}_t^N \subset \mathcal{F}_t$ (and we mean strict inclusion) then (i) is more restrictive.

Let N be an **F**-Poisson process. Then it is easy to realize that $N(t) - A(t)$ is an **F**-martingale. The following characterization result is the first important result linking point processes and martingales.

THEOREM 11. (Watanabe 1964). *A point process N is an **F**-Poisson process with intensity measure A if and only if $N(t) - A(t)$ is an **F**-martingale.*

REMARK 12. Definition 1 works well also if $A(\cdot)$ is not continuous. In the "martingale approach" the restriction to simple point processes is essential. If $0 < A(t) - A(t-) \leq 1$ then $N(t) - N(t-) = 1$ with probability $A(t) - A(t-)$, and thus N is *not* a Poisson process.
□

Let us now – for a moment – argue heuristically again. In some way $\hat{\Lambda}$ plays the same rôle for a general point process as A does for a Poisson process. From (7?) we "get"

$$E^{\mathcal{F}_s}[N(t) - \hat{\Lambda}(t)] = N(s) - \hat{\Lambda}(s) + E^{\mathcal{F}_s}\left[\int_s^t dN(y) - d\hat{\Lambda}(y)\right]$$

$$= N(s) - \hat{\Lambda}(s) + \int_s^t E^{\mathcal{F}_s}[E^{\mathcal{F}_y}[dN(y) - \hat{\lambda}(y)\,dy]] = N(s) - \hat{\Lambda}(s)$$

2.1 Models allowing for size fluctuation

and thus $N(t)-\hat{\Lambda}(t)$ ought to be an **F**-martingale. The way the to overcome the technical difficulties is to take this as definition.

DEFINITION 13. Let N be a point process and let $\hat{\Lambda}$ be a diffuse random measure. $\hat{\Lambda}$ is called the **F**-*compensator* of N if $\hat{\Lambda}$ is adapted to **F** and if $N(t) - \hat{\Lambda}(t)$ is an **F**-martingale.

REMARK 14. The compensator is uniquely (up to an equivalence) determined by the point process, and, conversely, the point process is uniquely determined by its compensator. With a certain extension of Definition 13 every point process has a compensator. The important restriction is that $\hat{\Lambda}$ is assumed to have continuous realizations. The compensator is continuous if and only if the jump times $S_1, S_2, \ldots$ are totally inaccessible stopping times. Roughly speaking this means that, although we follow the history of N, there does not exist a predictor which hits a future jump with positive probability. Further, $\hat{\Lambda}(\infty) = \infty$ P-a.s. if and only if $N(\infty) = \infty$ P-a.s.
□

Now there is no problem to define the inverse $\hat{\Lambda}^{-1}(t)$ and thus $\tilde{N} = N \circ \hat{\Lambda}^{-1}$ is well-defined. Further, $\hat{\Lambda}^{-1}(t)$ is an **F**-stopping time (Elliott 1982, pp. 67). Define the σ-algebra $\mathcal{G}_t$ by

$$\mathcal{G}_t = \mathcal{F}_{\hat{\Lambda}^{-1}(t)} = \sigma\{A \cap \{s \leq \hat{\Lambda}^{-1}(t)\};\ A \in \mathcal{F}_s,\ t \geq 0\}.$$

Since $\hat{\Lambda}^{-1}(\cdot)$ is increasing $\mathbf{G} = \{\mathcal{G}_t;\ t \geq 0\}$ is a filtration.

We now have the desired generalization of Lemma 4, which proves (12?), see Liptser and Shiryayev (1978, pp. 280 - 281).

THEOREM 15. Let N be a point process with **F**-compensator $\hat{\Lambda}$. If $N(\infty) = \infty$ P-a.s., then $\tilde{N} = N \circ \hat{\Lambda}^{-1}$ is a standard **G**-Poisson process.

Now we consider Cox processes.

DEFINITION 16. A point process N with **F**-compensator $\hat{\Lambda}$ is called an **F**-Cox process if

(i) $\hat{\Lambda}$ is $\mathcal{F}_0$-measurable.

(ii) $N(t) - N(s)$ is Poisson distributed with mean $\hat{\Lambda}(t) - \hat{\Lambda}(s)$ relative to $\mathcal{F}_s$.

REMARK 17. An **F**-Cox process where $\mathcal{F}_0$ is trivial, i.e., $\mathcal{F}_0 = \{\emptyset, \Omega\}$, has deterministic compensator and is thus an **F**-Poisson process.
□

The following proposition gives the relation between Cox processes (as in Definition 8) and **F**-Cox processes.

PROPOSITION 18. Let Λ be a diffuse random measure with $E[\Lambda(t)] < \infty$ for each $t < \infty$ and let **F** be given by $\mathcal{F}_t = \mathcal{F}_\infty^\Lambda \vee \mathcal{F}_t^N$. Then "Cox process corresponding to Λ" and "**F**-Cox process" are equivalent concepts.

The assumption $E[\Lambda(t)] < \infty$ is not necessary, and only due to our not quite general definition of compensators. Before proving Proposition 18 we shall give a simple lemma (Serfozo 1972, pp. 307 - 308).

LEMMA 19. *Let Λ be a random measure. N is the Cox process corresponding to Λ if and only if*

(i) $N(t)$ *has independent increments relative to $\mathcal{F}_\infty^\Lambda$;*

(ii) $N(t) - N(s)$ *is Poisson distributed with mean $\Lambda(t) - \Lambda(s)$ relative to $\mathcal{F}_\infty^\Lambda$.*

PROOF OF PROPOSITION 18: First we note that (i) and (ii) are equivalent to (ii) with "relative to $\mathcal{F}_\infty^\Lambda$" replaced by "relative to $\mathcal{F}_\infty^\Lambda \vee \mathcal{F}_s^N$."
Let N be a "Cox process corresponding to Λ." Then

$$E^{\mathcal{F}_\infty^\Lambda \vee \mathcal{F}_s^N}[N(t) - N(s)] = \Lambda(t) - \Lambda(s)$$

and thus, here we use the assumption $E[\Lambda(t)] < \infty$, see Definition 1.11 (ii), where Λ is the compensator of N. Since Λ is $\mathcal{F}_\infty^\Lambda$-measurable by definition it follows that N is an "F-Cox process."

If N is an "F-Cox process" it follows that N is a "Cox process corresponding to $\hat{\Lambda}$." Whether the underlying random measure is denoted by $\hat{\Lambda}$ or Λ is only a matter of notation. ∎

The following generalization of Watanabe's theorem (Theorem 11) is due to Grigelionis (1975, p. 105).

THEOREM 20. *A point process N with F-compensator $\hat{\Lambda}$ is an F-Cox process if and only if $\hat{\Lambda}$ is $\mathcal{F}_0$-measurable.*

□ □

Now let N be a point process with (continuous) F-compensator $\hat{\Lambda}$. Consider the risk process

$$X(t) = (1+\rho)\mu\hat{\Lambda}(t) - \sum_{k=1}^{N(t)} Z_k.$$

It follows from Theorem 15, and the above reasoning, that $\Psi(u) = \tilde{\Psi}(u)$, where $\tilde{\Psi}(u)$ is the ruin probability corresponding to a standard Poisson process with relative safety loading ρ.

NOTES: This result was, essentially, proved by De Vylder (1977). Although the pioneering work by Brémaud (1972) had appeared, the "martingale approach" to point processes was at that time not generally known. In fact, De Vylder (1977) gives intuitive arguments similar to ours, but states that "it seems impossible to define a new time depending stochastically on the real line." Instead De Vylder (1977) used an adaptation of the "martingale approach" discussed in Section 1.1. The underlying point process N was defined by conditional intensities, which essentially means that its $\mathbf{F}^N$-compensator was specified.

2.2 Models allowing for risk fluctuation

We shall now strictly keep ourselves to "risk fluctuations" which means that the gross risk premium c is *not* allowed to fluctuate. The purpose of this section is to discuss the choice of the point process describing the occurrence of the claims. The discussion will be based on the general theory of point processes, which – in contrast to the "martingale approach" to point processes – might be called the "random measure approach" to point processes. Standard references to that approach are Daley and Vere-Jones (1988), Matthes et al. (1978), Kallenberg (1983), and Karr (1986). We shall – at least later – rely much on Franken et al. (1981).

POINT PROCESSES AND RANDOM MEASURES

Although we have already discussed point processes slightly informally, we shall start with a number of basic definitions.

BASIC DEFINITIONS

The *phase space* is the space where the points (or claims) are located. In the risk model the phase space is $R_+ = [0, \infty)$. It is, however, more convenient to consider the phase space $R = (-\infty, \infty)$. A point process on R_+ is then interpreted as the restriction of a point process on R to R_+.

Let $\mathcal{B}(R)$ be the *Borel algebra* on R, i.e., the σ-algebra generated by open sets. A *Borel measure* μ on $(R, \mathcal{B}(R))$ is a (non-negative) measure such that its distribution function $\mu(t)$, $t \in R$, is finite. Formally the distribution function is defined by

$$\mu(t) = \begin{cases} \mu\{(0,t]\} & \text{if } t > 0 \\ 0 & \text{if } t = 0 \\ -\mu\{(t,0]\} & \text{if } t < 0 \end{cases}.$$

The same notation will be used for the measure and its distribution function. Let $\mathcal{M}$ denote the set of Borel measures.

DEFINITION 21. Let $\mu, \mu_1, \mu_2, \ldots \in \mathcal{M}$ be given. We say that μ_n converges *vaguely* to μ and write $\mu_n \to \mu$ if $\mu_n(t) \to \mu(t)$ for all $t \in R$ such that $\mu(\cdot)$ is continuous at t.

Endowed with the *vague topology*, i.e., the topology generated by vague convergence, $\mathcal{M}$ is a Polish space. Denote by $\mathcal{B}(\mathcal{M})$ the Borel algebra on $\mathcal{M}$. Further, $\mathcal{B}(\mathcal{M})$ equals the σ-algebra generated by projections, i.e., $\mathcal{B}(\mathcal{M}) = \sigma\{\mu(t) < y; \mu \in \mathcal{M}, t \in R, y < \infty\}$.

DEFINITION 22. A *random measure* is a measurable mapping from a probability space $(\Omega, \mathcal{F}, P)$ into $(\mathcal{M}, \mathcal{B}(\mathcal{M}))$.

Usually a random measure is denoted by Λ. The *distribution of* Λ is a probability measure Π on $(\mathcal{M}, \mathcal{B}(\mathcal{M}))$. Definitions 22 and 6 are, of course, equivalent.

DEFINITION 23. A random measure Λ is said to have *intensity* measure $E[\Lambda]$ if $E[\Lambda] \in \mathcal{M}$.

Obviously $E[\Lambda]$ – defined by $E[\Lambda]\{A\} = E[\Lambda\{A\}]$ for $A \in \mathcal{B}(\mathbf{R})$ – is a measure, but we shall only talk about the intensity when $E[\Lambda](t) = E[\Lambda(t)] < \infty$ for $t \in \mathbf{R}$.

Let $\Lambda, \Lambda_1, \Lambda_2, \ldots$ be random measures. As in the space D endowed with the Skorohod J_1 topology, we say that Λ_n converges in distribution to Λ and write $\Lambda_n \xrightarrow{d} \Lambda$, if $E[f(\Lambda)] \to E[f(\Lambda)]$ for all bounded and continuous real-valued functions f on $\mathcal{M}$. There is, however, one important difference. In D "convergence in distribution" is much more informative than "convergence of finite dimensional distributions," while convergence in distribution of Λ_n to Λ is equivalent to

$$(\Lambda_n\{I_1\}, \ldots, \Lambda_n\{I_k\}) \xrightarrow{d} (\Lambda\{I_1\}, \ldots, \Lambda\{I_k\})$$

for all k and all intervals $I_1, \ldots, I_k \subset \mathbf{R}$ with $P\{\Lambda\{\partial I_j\} = 0\} = 1$. (For an interval I the boundary ∂I is just the endpoints.)

Let $\mathcal{N} \in \mathcal{B}(\mathcal{M})$ denote the integer or infinite valued elements in $\mathcal{M}$. Endowed with the relative topology $\mathcal{N}$ is a Polish space and $\mathcal{B}(\mathcal{N})$ denotes the Borel algebra on $\mathcal{N}$. The elements in $\mathcal{N}$ are usually denoted by ν.

DEFINITION 24. A random measure with distribution Π is called a *point process* if $\Pi\{\mathcal{N}\} = 1$.

Put

$$\mathcal{N}_S = \{\nu \in \mathcal{N};\ \nu(t) - \nu(t-) = 0 \text{ or } 1\}$$

and

$$\mathcal{M}_D = \{\mu \in \mathcal{M};\ \mu(\cdot) \text{ is continuous}\}.$$

We have $\mathcal{N}_S \in \mathcal{B}(\mathcal{N})$ and $\mathcal{M}_D \in \mathcal{B}(\mathcal{M})$. Thus, cf. Remark 2 and Definition 7, a point process is simple if $\Pi\{\mathcal{N}_S\} = 1$ and a random measure is diffuse if $\Pi\{\mathcal{M}_D\} = 1$.

A point process is generally denoted by N. Any $\nu \in \mathcal{N}$ can be looked upon as a *realization* of N, and any $\nu \in \mathcal{N}_S$ as a realization of a simple point process. The epoch of the kth point of ν is denoted by $s_k(\nu)$. We always number the points such that

$$s_k(\nu) \leq s_{k+1}(\nu) \quad \text{and} \quad s_0(\nu) \leq 0 < s_1(\nu).$$

For $\nu \in \mathcal{N}_S$ we thus have

$$\cdots < s_{-1}(\nu) < s_0(\nu) \leq 0 < s_1(\nu) < s_2(\nu) < \cdots$$

When there is no risk for misunderstanding we denote the epoch of the kth point of a point process N by S_k instead of $s_k(N)$.

DEFINITION 25. The *shift operator* $T_x : \mathcal{M} \to \mathcal{M}$ is defined by
$$(T_x\mu)\{A\} = \mu\{A + x\} \qquad \text{for } A \in \mathcal{B}(\mathbf{R}) \text{ and } x \in \mathbf{R},$$
where $A + x = \{t \in \mathbf{R};\ t - x \in A\}$.

For any $B \in \mathcal{B}(\mathcal{M})$ we put $T_x B = \{\mu \in \mathcal{M};\ T_{-x}\mu \in B\}$.

DEFINITION 26. A random measure on $\mathbf{R}$ with distribution Π is called (strictly) *stationary* if $\Pi\{T_x B\} = \Pi\{B\}$ for all $x \in \mathbf{R}$ and all $B \in \mathcal{B}(\mathcal{M})$.

Definition 26 just means that a random measure Λ is stationary if the random process $\Lambda(t)$ has stationary increments. The intensity measure of a stationary random measure Λ is proportional to the Lebesgue measure, i.e., $E[\Lambda]\{dt\} = \alpha\, dt$. In this case α is called the *intensity*. The definition of stationarity is more natural on $\mathbf{R}$ than on $\mathbf{R}_+$ since $\mathbf{R}$ is, with respect to $+$, a group while $\mathbf{R}_+$ only is a semi-group. This is one of several reasons why we choose to consider $\mathbf{R}$ as the phase space.

To each $\mu \in \mathcal{M}$ there exists a Poisson process N with μ as its intensity measure. Let Π_μ denote its distribution. For each $B \in \mathcal{B}(\mathcal{N})$ the mapping $\mu \mapsto \Pi_\mu\{B\}$ is measurable (Grandell 1976, pp. 5 - 7). Thus to any random measure Λ with distribution Π the distribution Π_Λ of the corresponding Cox process is given by

$$\Pi_\Lambda\{B\} = \int_{\mathcal{M}} \Pi_\mu\{B\}\, \Pi\{d\mu\} \qquad \text{for } B \in \mathcal{B}(\mathcal{N}). \tag{13}$$

A Cox process is stationary if and only the underlying random measure is stationary.

DEFINITION 27. A Cox process with underlying random measure $\Lambda\{dt\} = \lambda \cdot dt$, where λ is a random variable, is called a *mixed Poisson process*.

REMARK 28. Actually a mixed Poisson process ought to be defined as a Cox process with $\Lambda = \lambda \cdot \mu$, for some $\mu \in \mathcal{M}$ and the process in Definition 27 ought to be called a "stationary mixed Poisson process." We do, however, only need stationary mixed Poisson processes.
□

Let N be a point process on $\mathbf{R}_+$ and let S_k denote the epoch of the kth point.

DEFINITION 29. N is called a *renewal process* (with inter-occurrence time distribution K^0) if the variables $S_1, S_2 - S_1, S_3 - S_2, \ldots$ are independent and if $S_2 - S_1, S_3 - S_2, \ldots$ have the same distribution K^0.

N is called an *ordinary* renewal process if S_1 also has distribution K^0. N is called a *stationary* renewal process if K^0 has finite mean $1/\alpha$ and if

S_1 has distribution K given by

$$K(t) = \alpha \int_0^t (1 - K^0(s))\, ds. \tag{14}$$

A stationary renewal process is a stationary point process in the sense of Definition 26. In a certain sense (14) holds for a general stationary point process and then the – now rather unmotivated – superscript 0 is natural.

SUPERPOSITION OF POINT PROCESSES

It is known that if many independent and "suitably" sparse point processes are added, then the sum is approximately a Poisson process.

We shall consider a null array

$$\begin{array}{llll} N_{1,1} & & & \\ N_{2,1} & N_{2,2} & & \\ \vdots & & \ddots & \\ N_{n,1} & N_{n,2} & \cdots & N_{n,n} \\ \vdots & & & \end{array}$$

of point processes such that $N_{n,1}, N_{n,2}, \ldots, N_{n,n}$ are independent for each n and further

$$\lim_{n \to \infty} \max_{1 \le k \le n} P\{N_{n,k}(t) \ge 1\} = 0.$$

Thus

$$N^{(n)} = \sum_{k=1}^n N_{n,k}$$

is a sum of point processes where each component becomes *asymptotically negligible*. This is, of course a condition of a certain sparseness of the individual components, but it is *not* enough to guarantee a Poisson limit. It holds, however, like in the case of random variables, that if $N^{(n)} \xrightarrow{d}$ some point process N as $n \to \infty$ then N is infinitely divisible.

DEFINITION 30. A point process N is called *infinitely divisible* if for each n there exists a point process N_n such that N has the same distribution as the sum of n indepentent copies of N_n.

A random measure Λ is called *infinitely divisible* if for each n there exists a random measure Λ_n such that Λ has the same distribution as the sum of n indepentent copies of Λ_n.

The theory of infinite divisibility is an important part of the general theory of point processes and is thoroughly discussed by Matthes et al. (1978) and Kallenberg (1983).

Let us go back to the question of "suitably" sparse point processes and Poisson limits. The following theorem is due to Grigelionis (1963).

THEOREM 31. *Let $\{N_{n,k}\}$ be a null array of point processes and let N be a Poisson process with intensity measure μ. Then $N^{(n)} \xrightarrow{d} N$ if and only if*

$$\lim_{n\to\infty} \sum_{k=1}^{n} P\{N_{n,k}(t) \geq 2\} = 0 \qquad \text{for all } t \in \mathbf{R}$$

and

$$\lim_{n\to\infty} \sum_{k=1}^{n} P\{N_{n,k}\{I\} \geq 1\} = \mu\{I\} \qquad \text{for all intervals } I \subset \mathbf{R}$$

with $\mu\{\partial I\} = 0$.

A Poisson process with intensity measure μ is infinitely divisible, and then N_n is a Poisson process with intensity measure μ/n. Similarly a Cox process is infinitely divisible if the underlying random measure is infinitely divisible. (There do, however, exist infinitely divisible Cox processes where the underlying random measure is *not* infinitely divisible.)

To our knowledge it is not known if there exist any infinitely divisible renewal processes except the Poisson process. A partial answer is, however, given by the following theorem due to Haberland (1976).

THEOREM 32. *Let N be a stationary renewal process with inter-occurrence time distribution K^0. If K^0 possesses a continuous derivative k^0 on $[0, \infty)$ (at 0 the right-hand derivative) then N is infinitely divisible if and only if it is a Poisson process.*

THINNING OF POINT PROCESSES

Let N be a point process with distribution Π, and denote by N_p the point process obtained by retaining (in the same location) every point of the process with probability p and deleting it with probability $q = 1 - p$, independently of everything else. N_p is called the *p-thinning* of N and N is called the *p-inverse* of N_p.

Let $\mathcal{P}$ be the set of distributions of point processes and let $\mathcal{C} \subset \mathcal{P}$ be the set of distributions of Cox processes.

Let $D_p : \mathcal{P} \to \mathcal{P}$ denote the p-thinning operator, i.e., for a point process N with distribution Π the distribution of the p-thinning of N is $D_p\Pi$. Put

$$\mathcal{D}_p = \{D_p\Pi;\ \Pi \in \mathcal{P}\},$$

i.e., $\mathcal{D}_p$ is the set of distributions of point processes, which can be obtained by p-thinning. The operator D_p is one to one, i.e., the *inverse* operator $D_p^{-1} : \mathcal{D}_p \to \mathcal{P}$ is unique. In order to simplify the notation we also allow the operators to act on the point processes themselves. Then $N_p = D_p N$ just means that N_p has distribution $D_p\Pi$.

For Poisson processes it is easy to realize that $D_p \Pi_\mu = \Pi_{p\mu}$ and conversely that $D_p^{-1} \Pi_\mu = \Pi_{\mu/p}$ for $p > 0$. Thus $\Pi_\mu \in \mathcal{D}_p$ for all $p > 0$. The same arguments hold for Cox processes, and thus $\mathcal{C} \subset \mathcal{D}_p$ for all p.

The following theorem, due to Kallenberg (1975), does not only contain almost everything known about "p-thinning" but also explains the importance of Cox processes in connection with thinning.

THEOREM 33. *Let $N_1, N_2, \ldots$ be point processes and let $p_1, p_2, \cdots \in (0,1)$ be real numbers such that $\lim_{n \to \infty} p_n = 0$. Then*

$$D_{p_n} N_n \xrightarrow{d} \text{some point process } N$$

if and only if

$$p_n N_n \xrightarrow{d} \text{some random measure } \Lambda.$$

If we have convergence it further holds that N is a Cox process with underlying random measure Λ.

In order to illustrate the power of Theorem 33, we shall derive Mecke's characterization of Cox processes as a corollary.

COROLLARY 34. (Mecke 1968). *A point process N with distribution Π can be obtained by p-thinning for every $p \in (0,1)$ if and only if it is a Cox process.*

PROOF: Formally Corollary 34 means that $\mathcal{C} = \cap_{0<p<1} \mathcal{D}_p$. Since we know that $\mathcal{C} \subset \cap_{0<p<1} \mathcal{D}_p$ it is enough to show that $\cap_{0<p<1} \mathcal{D}_p \subset \mathcal{C}$.

Let N be a point process with distribution $\Pi \in \cap_{0<p<1} \mathcal{D}_p$. Let $p_1, p_2, \ldots$ be as in Theorem 33 and let $N_1, N_2, \ldots$ be point processes with distributions $D_{p_1}^{-1}\Pi, D_{p_2}^{-1}\Pi, \ldots$. Since $D_{p_n} D_{p_n}^{-1} \Pi = \Pi$ we have, trivially, that $D_{p_n} N_n \xrightarrow{d} N$ and thus there exists a random measure Λ such that $p_n N_n \xrightarrow{d} \Lambda$. This implies that N is a Cox process with underlying random measure Λ. Thus $\cap_{0<p<1} \mathcal{D}_p \subset \mathcal{C}$ and the corollary follows. ∎

DEFINITION 35. *A point process N with distribution Π is called a top process if it cannot be obtained by p-thinning for any $p \in (0,1)$.*

Formally this means a point process N with distribution Π is a top process if $\Pi \notin \cup_{0<p<1} \mathcal{D}_p$. Obviously a deterministic point process is a top process, so top processes do exist.

Let N be a point process with distribution Π which is neither a Cox process nor a top process. Then there exists a unique value $p_N \in (0,1)$ such that

$$\Pi \begin{cases} \in \mathcal{D}_p & \text{if } p_N < p < 1 \\ \notin \mathcal{D}_p & \text{if } 0 < p < p_N \end{cases}.$$

Thedéen (1986) has given a theorem which says that $\Pi \in \mathcal{D}_{p_N}$, i.e., $D_{p_N}^{-1}\Pi$ is the distribution of a top process. Thus any point process is either a Cox process, a top process, or a thinning of a top process.

Now we consider renewal processes. In order to keep the presentation reasonably short we restrict ourselves to stationary renewal processes. Then K^0 has finite mean $1/\alpha$.

Kingman (1964) has characterized the class of renewal processes which are also Cox processes. Put

$$\hat{k}^0(v) = \int_0^\infty e^{-vs}\, dK^0(s).$$

THEOREM 36. (Kingman 1964). *A stationary renewal process is a Cox process if and only if*

$$\hat{k}^0(v) = \frac{1}{1 + bv + \int_0^\infty (1 - e^{-vx})\, dB(x)} \tag{15}$$

for some $b \geq 0$ and some measure B on $(0, \infty)$ such that $\int_0^\infty x\, dB(x) < \infty$.

Let us now consider a point process N which is both a renewal process and a Cox process in terms of the underlying random measure. Put

$$c = \int_0^\infty dB(x) \quad \text{and} \quad C(x) = B(x)/c \text{ if } c < \infty. \tag{16}$$

$b = 0$ *and* $c < \infty$.
In this case, see Grandell (1976, p. 39),

$$\Lambda(t) = \sum_{k=1}^{\tilde{N}(t)} \ell_k, \quad \left(\sum_{k=1}^0 \ell_k = 0\right),$$

where $\tilde{N}$ is a stationary renewal process with inter-occurrence time distribution C and $\{\ell_k\}_{k=1}^\infty$ is a sequence, independent of $\tilde{N}$, of independent exponentially distributed random variables with mean $1/c$. This is the only case where N is not simple.

$b = 0$ *and* $c = \infty$.
In this case, see Grandell (1976, p. 39), Λ is diffuse but P-a.s. singular with respect to the Lebesgue measure. Surprisingly this will turn out to be an interesting case.

$b > 0$ *and* $c = 0$.
In this case N is a Poisson process with intensity $1/b$.

$b > 0$ *and* $c > 0$.
In this case Λ has the representation

$$\Lambda(t) = \int_0^t \lambda(s)\, ds$$

and $\lambda(s)$ alternates between the values 0 and $1/b$ in such a way that $\lambda(s)$ is proportional to a stationary "regenerative phenomenon" (cf. Kingman 1972, p. 48).

$b > 0$ and $0 < c < \infty$.

(This case was formally included in the above case, but now the "regenerative phenomenon" has a simple interpretation.) In this case $\lambda(s)$ alternates between the values 0 and $1/b$ on intervals whose lengths are independent random variables. The lengths of the intervals where $\lambda(s) = 1/b$ are exponentially distributed with mean b/c and the lengths of the intervals where $\lambda(s) = 0$ have distribution C.

EXAMPLE 37. From the point of view of Cox processes the case $b > 0$ and $0 < c < \infty$ is the most interesting one. This is also especially true when the lengths of the intervals where $\lambda(s) = 0$ are exponentially distributed, since then $\lambda(s)$ is a *two-state Markov process*. We shall consider two-state Markov processes and generally we denote the two states by α_1 and α_2 and the mean times in the states by $1/\eta_1$ and $1/\eta_2$, respectively. Note that η_k is the *intensity* of the exponential distribution corresponding to the time in state α_k. Thus we have

$$\alpha_1 = 0, \qquad \alpha_2 = \frac{1}{b}, \qquad \frac{1}{\eta_1} = \int_0^\infty x\, dC(x), \quad \text{and} \quad \eta_2 = \frac{c}{b}.$$

Using these notations it follows from (15) that

$$\hat{k}^0(v) = \frac{1}{1 + \frac{v}{\alpha_2} + \frac{\eta_2}{\alpha_2}\left(1 - \frac{\eta_1}{\eta_1+v}\right)} = \frac{\alpha_2\eta_1 + \alpha_2 v}{v^2 + v(\alpha_2 + \eta_1 + \eta_2) + \alpha_2\eta_1}.$$

Let $-\theta_1$ and $-\theta_2$ denote the solutions of $v^2 + v(\alpha_2 + \eta_1 + \eta_2) + \alpha_2\eta_1 = 0$, i.e.,

$$\theta_1 = \frac{1}{2}(\alpha_2 + \eta_1 + \eta_2) - \sqrt{\frac{1}{4}(\alpha_2 + \eta_1 + \eta_2)^2 - \alpha_2\eta_1},$$

$$\theta_2 = \frac{1}{2}(\alpha_2 + \eta_1 + \eta_2) + \sqrt{\frac{1}{4}(\alpha_2 + \eta_1 + \eta_2)^2 - \alpha_2\eta_1}.$$

Thus we have

$$\hat{k}^0(v) = \frac{\theta_1\theta_2 + \alpha_2 v}{(v+\theta_1)(v+\theta_2)}$$

$$= \frac{\theta_2 - \alpha_2}{\theta_2 - \theta_1}\frac{\theta_1}{\theta_1 + v} + \left(1 - \frac{\theta_2 - \alpha_2}{\theta_2 - \theta_1}\right)\frac{\theta_2}{\theta_2 + v} \qquad (17)$$

or

$$K^0(t) = \frac{\theta_2 - \alpha_2}{\theta_2 - \theta_1}(1 - e^{-\theta_1 t}) + \left(1 - \frac{\theta_2 - \alpha_2}{\theta_2 - \theta_1}\right)(1 - e^{-\theta_2 t}). \qquad (18)$$

Thus $K^0(t)$ is a mixture of exponential distributions.
□

Let N be a stationary renewal process with inter-occurrence time distribution K^0. It is often rather difficult to check if $\hat{k}^0$ has the representation (15). The following theorem, due to Thorin (1988), shows that a wide class of renewal processes are, in fact, Cox processes. The proof, based on Stieltjes transforms, is inspired by Berg (1981).

THEOREM 38. *A stationary renewal process with*

$$K^0(t) = \int_0^\infty (1 - e^{-t\theta}) \, dV(\theta),$$

where V is a distribution function with $\int_0^\infty \frac{1}{\theta} \, dV(\theta) < \infty$, is a Cox process.

PROOF: A function $f : (0, \infty) \to \mathbf{R}_+$ is called a *Stieltjes transform* if

$$f(v) = a + \int_0^\infty \frac{dA(s)}{v+s}$$

for some $a \geq 0$ and some Borel measure A on $\mathbf{R}_+$ such that

$$\int_1^\infty \frac{1}{s} \, dA(s) < \infty.$$

The pair (a, A) is determined by f, and we especially have $\lim_{v \to 0} v f(v) = A\{\{0\}\}$. Reuter (1956) has shown that if $f \not\equiv 0$ then $1/(vf(v))$ is also a Stieltjes transform.

Now we have

$$\hat{k}^0(v) = \int_0^\infty \int_0^\infty \theta e^{-t(v+\theta)} \, dt \, dV(\theta) = \int_0^\infty \frac{\theta \, dV(\theta)}{v+\theta}$$

and thus $\hat{k}^0(v)$ is a Stieltjes transform. Thus, for some pair (a, A),

$$\frac{1}{\hat{k}^0(v)} = av + v \int_0^\infty \frac{dA(s)}{v+s} = av + A\{\{0\}\} + \int_{0+}^\infty \frac{v \, dA(s)}{v+s}.$$

Since

$$A\{\{0\}\} = \lim_{v \to 0} \frac{v}{v\hat{k}^0(v)} = \lim_{v \to 0} \frac{1}{\hat{k}^0(v)} = 1$$

we get

$$\frac{1}{\hat{k}^0(v)} = 1 + av + \int_{0+}^\infty \frac{v \, dA(s)}{v+s} = 1 + av + \int_{0+}^\infty \frac{v+s-s}{v+s} \, dA(s)$$

$$= 1 + av + \int_{0+}^\infty \left(\frac{1}{s} - \frac{1}{v+s}\right) s \, dA(s)$$

$$= 1 + av + \int_{0+}^\infty s \, dA(s) \int_0^\infty \left(e^{-sx} - e^{-(v+s)x}\right) dx$$

$$= 1 + av + \int_0^\infty (1 - e^{-vx}) \int_{0+}^\infty se^{-sx} \, dA(s) \, dx.$$

Let the measure B on $(0, \infty)$ be defined by

$$dB(x) = \int_{0+}^\infty se^{-sx} \, dA(s) \, dx.$$

Using

$$\infty > \frac{1}{\alpha} = \lim_{v \to 0} \left(\frac{1}{\hat{k}^0(v)} - 1\right) = a + \int_{0+}^\infty \frac{dA(s)}{s}$$

we get
$$\int_0^\infty x\,dB(x) = \int_{0+}^\infty \frac{dA(s)}{s} < \infty$$
and the theorem follows from Theorem 36. ∎

General mixtures of exponential distributions were discussed in Section 1.3 as claim distributions. From Theorem 38 and that discussion it then follows that stationary renewal processes with Pareto- and (certain) Γ-distributed inter-occurrence times are Cox processes. In Example 42 we consider the Γ-case in some detail. It follows from the proof of Theorem 38 that the measure B is also a mixture of exponential distributions, although A does not need to be a probability measure.

If B is a mixture of exponential distributions if follows from the proof of Theorem 38 – in the opposite direction – that K^0 is also a mixture of exponential distributions. If B is a finite mixture of order n, i.e., if

$$B(x) = c \cdot \sum_{k=1}^n (1 - e^{-x\theta_k}) p_k,$$

where $\{p_k\}$ is a probability distribution, we get from (15) that

$$\hat{k}^0(v) = \frac{1}{1 + bv + c\bigl(1 - \sum_{k=1}^n \frac{\theta_k}{\theta_k + v} p_k\bigr)}.$$

By partial fraction decomposition it follows that K^0 is a finite mixture of order $n+1$. The exact relation between B and K^0 is, however, complicated, as seen from Example 37.

Now we consider thinning of renewal processes. A p-thinned renewal process is, of course, a renewal process and we allow the operator D_p to act on K^0 and $\hat{k}^0$, i.e., $D_p K^0$ is the inter-occurrence time distribution of the p-thinned process and $D_p \hat{k}^0$ is its Laplace transform. Then

$$D_p \hat{k}^0(v) = \sum_{j=1}^\infty q^{k-1} p \hat{k}^0(v)^j = \frac{p\hat{k}^0(v)}{1 - q\hat{k}^0(v)} \tag{19}$$

and conversely

$$\hat{k}^0(v) = \frac{D_p \hat{k}^0(v)}{p + q D_p \hat{k}^0(v)}. \tag{20}$$

Yannaros (1988b) has shown that renewal processes can only be obtained by p-thinning of renewal processes. This means that if N is a renewal process, then the corresponding top process must also be a renewal process.

The following "theorem" is almost a triviality.

THEOREM 39. *A stationary renewal process is:*

(i)   a Cox process if and only if $\frac{\hat{k}^0(v)}{p + (1-p)\hat{k}^0(v)}$ is a Laplace transform for all $p \in (0,1)$;

(ii)   a top process if and only if $\frac{\hat{k}^0(v)}{p+(1-p)\hat{k}^0(v)}$ is not a Laplace transform for any $p \in (0,1)$.

In spite of its triviality, it may sometimes be easier to check condition (i) than to check if $\hat{k}^0$ has the representation (15) or if Theorem 38 is applicable.

Little seems to be known about characterization of top processes. The following theorem is sometimes useful.

THEOREM 40. (Yannaros 1988b). *A stationary renewal process is a top process if*

$$\lim_{t\to\infty} \frac{1 - K^{0^{2*}}(t)}{t[1 - K^0(t)]} = \infty, \tag{21}$$

where $K^{0^{2*}}$ is the convolution of K^0 with itself. If K^0 has density k^0, then (21) holds if

$$\lim_{t\to\infty} \frac{k^{0^{2*}}(t)}{tk^0(t)} = \infty. \tag{22}$$

INDICATION OF PROOF: The proof is based on the following inequality, due to Svensson (1987).

Let N_p be a p-thinning of N and let $A \in \mathcal{B}(\mathbf{R}_+)$ such that $E[N_p\{A\}] < \infty$ be given. Then

$$P\{N_p\{A\} = 0\} = \sum_{j=0}^{\infty} (1-p)^j P\{N\{A\} = j\} = E\left[(1-p)^{N\{A\}}\right]$$

and

$$P\{N_p\{A\} = 1\} = \sum_{j=0}^{\infty} jp(1-p)^{j-1} P\{N\{A\} = j\}$$

$$= \frac{p}{1-p} E\left[N\{A\}(1-p)^{N\{A\}}\right].$$

Since $E[N_p\{A\}] = pE[N\{A\}]$ and since $N\{A\}$ and $(1-p)^{N\{A\}}$ are negatively correlated we get

$$P\{N_p\{A\} = 1\} \leq \frac{p}{1-p} E[N\{A\}] \cdot E\left[(1-p)^{N\{A\}}\right]$$

$$= \frac{1}{1-p} E[N_p\{A\}] \cdot P\{N_p\{A\} = 0\}.$$

Thus

$$\frac{P\{N_p\{A\} = 1\}}{P\{N_p\{A\} = 0\} E[N_p\{A\}]} \leq \frac{1}{1-p},$$

which is Svensson's inequality. (If $E[N_p\{A\}] = \infty$ the inequality is trivially true.)

The theorem follows easily from the inequality with $A = (0,t]$ and the fact that renewal processes can only be obtained by p-thinning of renewal processes. ∎

REMARK 41. From the proof of Theorem 40 it follows that
$$\sup_{0<t<\infty} \frac{1 - K^{0^{2*}}(t)}{t[1 - K^0(t)]} = \infty$$
implies a top process. If the distribution K^0 has bounded support, i.e., if there exists $t_0 < \infty$ such that $K^0(t) < 1$ for $t < t_0$ and $K^0(t) = 1$ for $t \geq t_0$, then the corresponding renewal process is a top process. □

EXAMPLE 42. (Yannaros). In most applications when generalizations of the Poisson process – in the direction of renewal processes – are of interest, the first generalization which comes into mind is a renewal process N with Γ-distributed inter-occurrence times. In that case we have
$$k^0(t) = \frac{t^{(\gamma-1)}}{\Gamma(\gamma)} e^{-t} \quad \text{and} \quad \hat{k}^0(v) = \frac{1}{(1+v)^\gamma}, \qquad (23)$$
where γ is called the form parameter and where we – for simplicity – have put the scale parameter equal to 1. For a random variable S with this distribution, it is well-known that $E[S] = \text{Var}[S] = \gamma$. Further, if γ is an integer, S has the same distribution as the sum of γ independent and exponentially distributed random variables. Those Γ-distributions are often referred to as Erlang distributions.

Yannaros (1988a) has shown that N is:

(i)   a Cox process if $0 < \gamma \leq 1$;

(ii)   a top process if $\gamma > 1$.

This result is – in our opinion – very interesting since it concerns an important renewal process and since it illustrates that transition between the "extreme" classes of Cox and top processes is not "continuous." Further it was very surprising – at least to the author – that such a simple renewal process can be a Cox process. Note that the Cox process cannot be in the – for Cox processes – natural class "$b > 0$ and $0 < c < \infty$." Therefore we shall consider this example in some detail.

$0 < \gamma < 1$.

This case follows from Theorem 38, since Thorin (1973) has shown that these Γ-distributions are mixtures of exponential distributions.

Due to the importance of this example we will – following Yannaros (1988a) – give an elementary derivation. We then rely on two well-known results for Laplace transforms, see Feller (1971, pp. 439 and 441).

A function $\varphi(v)$ on $(0, \infty)$ is a Laplace transform of a probability distribution if and only if $\varphi(0+) = 1$ and φ is completely monotone, i.e., $(-1)^n \varphi^{(n)}(v) \geq 0$.

If φ is completely monotone and ψ a positive function with a completely monotone derivative then $\varphi(\psi)$ is completely monotone.

It follows from Theorem 39 (i) it is enough to show that

$$D_p^{-1}\hat{k}^0(v) = \frac{(1+v)^{-\gamma}}{p+(1-p)(1+v)^{-\gamma}} = \frac{1}{1-p+p(1+v)^\gamma}$$

is completely monotone since $D_p^{-1}\hat{k}^0(0) = 1$. We can write

$$D_p^{-1}\hat{k}^0(v) = \varphi(\psi(v)), \quad \text{where} \quad \varphi(v) = \frac{1}{1-p+pv} \quad \text{and} \quad \psi(v) = (1+v)^\gamma.$$

The function φ is proportional to the Laplace transform of an exponential distribution, and thus it is completely monotone, $\psi(v) > 0$, and $\psi'(v) = \gamma(1+v)^{(\gamma-1)} > 0$. Further, $\gamma - n < 0$ for all $n \geq 1$. At each derivation of $\psi'(v)$ we thus have a change of signs and it follows that $\psi'(v)$ is completely monotone. Thus $D_p^{-1}\hat{k}^0(v)$ is a Laplace transform.

It follows from Theorem 36 that

$$(1+v)^\gamma = 1 + bv + \int_0^\infty (1-e^{-vx})\, dB(x) > bv$$

and thus $b < \frac{(1+v)^\gamma}{v} \to 0$ as $v \to \infty$ which implies that $b = 0$. Since N is simple we must have $c = \infty$. Thus Λ is diffuse but P-a.s. singular with respect to the Lebesgue measure.

$\gamma = 1$.
Since the inter-occurrence times are exponentially distributed, N is in fact a standard Poisson process.

$\gamma > 1$.
In this case

$$k^{0^{2*}}(t) = \frac{t^{(2\gamma-1)}}{\Gamma(2\gamma)} e^{-t} \quad \text{and thus} \quad \frac{k^{0^{2*}}(t)}{tk^0(t)} = t^{(\gamma-1)}\frac{\Gamma(\gamma)}{\Gamma(2\gamma)} \to \infty$$

as $t \to \infty$. Thus it follows from Theorem 40 that N is a top process.

This result was first proved by Yannaros (1985) in the case $\gamma = 2, 3, \ldots$ and generalized to arbitrary $\gamma > 0$ by Kolsrud (1986). Both these proofs are quite different from the one given here.

The claims faced by an insurance company is, of course, the sum of all claims caused by the policyholders. To policyholder number k we can associate an *individual* point process N_k which describes the epochs of the claims of that policyholder. In pure life insurance one "claim" can occur at most, namely the death of the policyholder. Thus $N_k(t)$ is equal to 0 or 1. In non-life insurance the individual point processes may be more complicated.

2 Generalizations of the classical risk model

Assume now that the individual point processes $N_1, N_2, \ldots$ are independent. The point process N is thus the sum of these individual point processes. Relying on Theorem 31 it then seems natural to assume that N is a Poisson process with some intensity measure μ. If we can disregard seasonal variation and other kinds of temporal variation, and if the variation of the number of policyholders involved in the portfolio is taken care of in the individual point processes, it is natural to put μ proportional to the Lebesgue measure. Thus we have a motivation for the classical risk model.

In some cases we may have a *"direct"* dependence between the individual point processes. With "direct" dependence we mean that a claim in one individual point process causes claims, or affects the probability of claims, in other individual point processes. As examples we may think of contagion and accidents in life and sickness insurance, the spread of fire to several buildings in fire insurance, and so on. We shall soon give an argument, where nothing is assumed about how N is built up by individual point processes.

Another kind of dependence may be called *"indirect"* dependence. We then think of cases where the whole risk situation may vary with variations in the environment. In, for example, automobile insurance important parts of the environment are weather conditions and traffic volume. If the individual point processes are independent conditioned upon the environment, we can again rely on Theorem 31 and it seems natural to assume that N is a Cox process. This is the reasoning we had in mind when we claimed that Cox processes are very natural as models for "risk fluctuation."

Consider now pure life insurance, where the only random quantity is the time of death of the policyholder. In a rich country like Sweden few deaths are directly caused by infectious deseases. This is, at least now, still true if we take AIDS into account. Also big accidents, like plane and train accidents, cause few deaths compared to the total number of deaths. Thus the direct dependence between the individual point processes seems to be almost negligible. Further, there is no famine and – more due to the geographical position – no serious nature catastrophes. Thus the indirect dependence between the individual point processes also seems to be almost negligible. (We have consciously disregarded armed conflicts and wars, since those probably cannot be taken into account in a model. At least in a war the solvency of insurance companies is a minor problem.) Finally, we disregard from possible "seasonal" variation in the death frequency. Thus the classical model seems to work well for pure life insurance in rich countries. On the other hand, risk theoretic considerations are probably not too interesting in this case, since fluctuations in the interest and other economical variation are more important to the insurance company than the random variation of the risk business.

Now we consider N but make no assumptions about the individual point processes and the relation between them. We shall now exploit an idea

2.2 Models allowing for risk fluctuation

which goes back to Almer (1957) and consider claims as caused by "risk situations" or *incidents*. To each incident we associate a *claim probability* p and we assume that incidents become claims independent of each other. Under these assumptions the point process describing the incidents is the p-inverse of the "claim process" N and will therefore be denoted by $D_p^{-1} N$. A rather general and realistic way to apply these ideas is to let the incidents be the "claims" in a population and p the proportion of the population insured in the insurance company under consideration. This indicates that *it is highly unnatural to choose N among top processes*.

Anyone who has driven a car has certainly experienced incidents and, hopefully, only few of them have resulted in accidents. This is probably the every day use of the word "incident." Let us therefore again consider automobile insurance. Suppose we can specify the concept "incident" and a claim probability p. One problem is that the incidents must be so generally defined that all, or at least almost all, claims can be associated with an incident. Generally this means that p will be small. In principle we may have a series of definitions of "incidents" and a corresponding series of probabilities $p_1, p_2, \ldots$ such that $\lim_{n \to \infty} p_n = 0$. Then it "follows" that N is a Cox process. Certainly this argument is very speculative and must not be taken too seriously. If every overtaking, every braking, every curve, and so on is regarded as an incident we may look upon the "incident process" more like an intensity than a point process. Then it is highly reasonable to believe that the claim probability p depends on the environment and we are back in the reasoning about "risk fluctuation." In spite of all reservations, this "incident" argument, in our opinion, indicates that *it is natural to choose N among Cox processes*.

From an analytical point of view it is natural to generalize the classical risk model to the "renewal model," i.e., where the occurrences of the claims are described by a renewal process. A natural characteristic of the inter-occurrence time distribution is the coefficient of variation CV, defined by

$$CV = \frac{\text{standard deviation}}{\text{mean}}.$$

From Example 42 it follows that a renewal process with Γ-distributed inter-occurrence times is

and
$$\text{a Cox process if } CV \geq 1$$
$$\text{a top process if } CV < 1.$$

Thus, by our arguments, the use of a Γ-renewal process with $CV \geq 1$ might be natural. Its representation as a Cox process gives, however, no information if, or when, it is a reasonable model.

Another possible choice, which has been used, is to let K^0 be a mixture of exponential distributions, i.e., $K^0(s) = \sum_{k=1}^n (1 - e^{-\theta_k s}) p_k$ where $p_k \geq 0$ and $\sum_{k=1}^n p_k = 1$. From Theorem 38 it follows that this renewal process can be represented as a Cox process. Certainly it has been used because of its simplicity, and not because of its relation to Cox processes. It corresponds to a Cox process where the intensity process $\lambda(t)$ alternates between the values 0 and α_2 in such a way that $\lambda(t)$ is a two-state Markov process. We do believe that Cox processes corresponding to two-state Markov processes are of interest. We shall consider them later, but then we let $\lambda(t)$ alternate beween two-states α_1 and α_2 where $\alpha_1 > 0$ is allowed. In our opinion, it is rather difficult to find situations where $\alpha_1 = 0$ is natural.

Another of our arguments is that it sometimes might be natural to consider N as the sum of independent individual point processes. Relying on Theorem 31 we used this as an argument to choose N as a Poisson process. If we only assume that N is a sum of independent asymptotically negligible point processes we can only draw the conclusion that N must be infinitely divisible. From Theorem 32 it follows that – essentially – the Poisson process is the only infinitely divisible renewal process.

Putting all this together, we do not find it very convincing that the occurrence of claims can be much more realistically described by renewal processes than with Poisson processes. This, however, does not mean that it is uninteresting to consider renewal models. One practical aspect is that we might be interested in whether ruin occurs only when it can be observed. In cases where the risk process is regularly observed we may want to consider a renewal model where K^0 is an one-point distribution, although the occurrences of the claims are described by a Poisson process. Then the ordinary renewal process is purely deterministic, and the "claims" are the "arrivals of the accountant." In our opinion, a much more important reason is mathematical clarity. By explicit use of an inter-occurrence time distribution, a better insight is achieved in how the ruin probability and the Lundberg exponent depend on the risk process.

CHAPTER 3
Renewal models

We shall now consider the case where the occurrence of the claims is described by a renewal process N. Let S_k denote the epoch of the kth claim. Recall from Section 2.2 that a point process on $\mathbf{R}_+$ is called a *renewal process* (with inter-occurrence time distribution K^0) if the variables S_1, $S_2 - S_1$, $S_3 - S_2, \ldots$ are independent and if $S_2 - S_1$, $S_3 - S_2, \ldots$ have the same distribution K^0. Further, N is called an *ordinary* renewal process if S_1 also has distribution K^0. N is called a *stationary* renewal process if K^0 has finite mean $1/\alpha$ and if S_1 has distribution K given by

$$K(t) = \alpha \int_0^t (1 - K^0(s))\, ds. \qquad (1)$$

The first treatment of the ruin problem when the occurrence of the claims is described by a renewal process N, is due to Sparre Andersen (1957). After the publication of his paper this model has been considered in several works. In a series of papers Thorin has carried through a systematic study based on Wiener-Hopf methods. Good references are Thorin (1974) and the review by Thorin (1982). Following Thorin we first consider the ordinary case.

3.1 Ordinary renewal models

Let N be an ordinary renewal process and assume that K^0 has finite mean $1/\alpha$. N is not stationary, unless K^0 is an exponential distribution, and $E[N(t)] \neq \alpha t$. The first problem is to define the relative safety loading. Therefore we consider the random variables X_k, $k = 1, 2, \ldots$, defined by

$$X_k \stackrel{\text{def}}{=} -[X(S_k) - X(S_{k-1})] = Z_k - c(S_k - S_{k-1}) \qquad (S_0 \stackrel{\text{def}}{=} 0). \qquad (2)$$

Obviously $X_1, X_2, \ldots$ is a sequence of independent and identically distributed random variables. This observation will be fundamental in the analysis. The expected loss between two claims is

$$E[X_k] = E[X_1] = -E[X(S_1)] = E[Z_1 - cS_1] = \mu - \frac{c}{\alpha} \quad (3)$$

and it is natural to define the relative safety loading ρ by

$$\rho = \frac{\frac{c}{\alpha} - \mu}{\mu} = \frac{c - \alpha\mu}{\alpha\mu} = \frac{c}{\alpha\mu} - 1$$

which is formally the same definition as in the stationary case. This is very natural since the only difference between the ordinary and the stationary case is the distribution of S_1. When nothing else is said we assume positive safety loading, i.e., $\rho > 0$.

Define the random variables Y_n, $n = 0, 1, 2, \ldots$, by

$$Y_0 = 0 \quad \text{and} \quad Y_n = \sum_{k=1}^{n} X_k \quad \text{for } n = 1, 2, \ldots \quad (4)$$

and note that $Y_n = -X(S_n)$. Y_n is thus the loss immediately after the nth claim. The ruin probability $\Psi^0(u)$, where the superscript 0 refers to the "ordinary case," is as always defined by

$$\Psi^0(u) = P\{u + X(t) < 0 \text{ for some } t > 0\}.$$

Since $c > 0$ ruin can only occur at claim epochs, we have

$$\Psi^0(u) = P\{\max_{n \geq 1} Y_n > u\}.$$

Let G denote the distribution function of X_n, i.e., $G(x) = P\{X_n \leq x\}$. Put

$$\gamma \stackrel{\text{def}}{=} E[X_n] = -\mu\rho < 0 \quad (5)$$

and

$$g(r) \stackrel{\text{def}}{=} \int_{-\infty}^{\infty} e^{rx} \, dG(x) = E[e^{rX_1}] = E[e^{r(Z_1 - cS_1)}] = (h(r) + 1)\hat{k}^0(cr), \quad (6)$$

where $h(r)$ is given by Definition 1.3, and where $\hat{k}^0(v) = \int_0^\infty e^{-vs} \, dK^0(s)$.

ASSUMPTION 1. $G(0) < 1$.

REMARK 2. The case $G(0) = 1$ is formally possible – take for example $S_n - S_{n-1} = 1/\alpha$ and $Z_n = \mu$ P-a.s. – but uninteresting, since it implies that $Z_n \leq c(S_n - S_{n-1})$ P-a.s. and thus $\Psi^0(u) \equiv 0$. In spite of its triviality it shows that (I) does *not* necessarily hold in the ordinary case. □

The function $g(r)$ will be important. From Assumption 1.4 it follows that $g(0) = 1$, $g'(0) = -\mu\rho < 0$, and that g is convex and continuous on $[0, r_\infty)$. Further $g(r) \to \infty$ when $r \uparrow r_\infty$. For $r_\infty < \infty$ it is obvious since $\hat{k}^0(cr_\infty) > 0$. If $r_\infty = \infty$ Assumption 1 must be used. Since G is right-continuous there exists $x_0 > 0$ such that $G(x_0) < 1$, and thus

$$g(r) \geq e^{rx_0}(1 - G(x_0)) \to \infty \quad \text{as } r \to \infty.$$

From this argument it follows that Definition 3 is, at least mathematically, meaningful.

DEFINITION 3. The Lundberg exponent R is in the renewal case the positive solution of
$$g(r) = 1. \tag{7}$$

REMARK 4. If S_1 is exponentially distributed with mean $1/\alpha$ we have
$$\hat{k}^0(v) = \frac{1}{1 + v/\alpha}$$
and thus
$$1 = g(R) = \frac{h(R) + 1}{1 + cR/\alpha} \quad \text{or} \quad h(R) = \frac{cR}{\alpha}.$$

Thus, see (1.12), the definition of R in the Poisson case is included in Definition 3.
□

The process $Y = \{Y_n; \ n = 0, 1, 2, \ldots\}$ is a *random walk*. A random walk can be looked upon as the discrete time correspondence of a continuous time process with stationary and independent increments. In the classical risk model the risk process was a somewhat special process with stationary and independent increments. In the "martingale approach" those special properties were not used, and it is therefore not surprising – as we shall see – that the derivation goes through almost word for word.

Consider the filtration $\mathbf{F}^Y = (\mathcal{F}_n^Y; \ n = 0, 1, 2, \ldots)$ where
$$\mathcal{F}_n^Y = \sigma\{Y_k; \ k = 0, \ldots, n\}.$$

Let N_u be the number of the claim causing ruin, i.e.,
$$N_u = \min\{n \mid Y_n > u\}.$$

As in the continuous time case N_u is a stopping time and
$$\Psi^0(u) = P\{N_u < \infty\}.$$

Put
$$M_u(n) = \frac{e^{-r(u-Y_n)}}{g(r)^n}.$$

REMARK 5. If we compare $M_u(n)$ with $M_u(t)$ as defined in Section 1.1 we observe the change of sign in the exponent. This is, of course, due to the fact that $Y_n = -X(S_n)$. The reader may be irritated by our, for the moment, quite unnecessary "change of sign." The reason is purely notational, and the choice is made in order to make the future application of random walk results easier.
□

Exactly as in (1.17) it follows that $M_u(n)$ is a martingale. Because of the "change of sign" we do, however, repeat the derivation. We have

$$E^{\mathcal{F}_k^Y}[M_u(n)] = E^{\mathcal{F}_k^Y}\left[\frac{e^{-r(u-Y_n)}}{g(r)^n}\right] = E^{\mathcal{F}_k^Y}\left[\frac{e^{-r(u-Y_k)}}{g(r)^k} \cdot \frac{e^{r(Y_n-Y_k)}}{g(r)^{(n-k)}}\right]$$

$$= M_u(k) \cdot E^{\mathcal{F}_k^Y}\left[\frac{e^{r(Y_n-Y_k)}}{g(r)^{(n-k)}}\right] = M_u(k).$$

Choose $n_0 < \infty$ and consider $n_0 \wedge N_u$ which is a bounded $\mathbf{F}^Y$-stopping time. Since Theorem 1.14 also holds in this case, it follows from (1.18) that

$$e^{-ru} = M_u(0) = E[M_u(n_0 \wedge N_u)] \geq E[M_u(N_u) \mid N_u \leq n_0]P\{N_u \leq n_0\}$$

and thus, since $u - Y_{N_u} \leq 0$ on $\{N_u < \infty\}$,

$$P\{N_u \leq n_0\} \leq \frac{e^{-ru}}{E[M_u(N_u) \mid N_u \leq n_0]}$$

$$\leq \frac{e^{-ru}}{E[g(r)^{-N_u} \mid N_u \leq n_0]} \leq e^{-ru} \max_{0 \leq n \leq n_0} g(r)^n.$$

When $n_0 \to \infty$ we get

$$\Psi^0(u) \leq e^{-ru} \sup_{n \geq 0} g(r)^n.$$

The best choice of r is the Lundberg exponent R, see (1.21). Thus we have Lundberg's inequality

$$\Psi^0(u) \leq e^{-Ru}. \tag{8}$$

Lundberg's inequality in the ordinary renewal case was first proved by Sparre Andersen (1957, p. 224) by completely different methods.

With exactly the same arguments as in the derivation of (1.23) we get

$$\Psi^0(u) = \frac{e^{-Ru}}{E[e^{-R(u-Y_{N_u})} \mid N_u < \infty]}. \tag{9}$$

EXAMPLE 6. EXPONENTIALLY DISTRIBUTED CLAIMS. Consider the case when Z_k is exponentially distributed. In the classical model we could prove (II) within the "martingale approach" by using the "lack of memory" of the exponential distribution. In order to handle the conditioning on $T_u < \infty$ we introduced the σ-algebra "strictly prior" to ruin $\mathcal{F}_{T_u-}^X$. This σ-algebra seems to have no natural correspondence in the discrete time case. However, in Section A.2 a – more complicated – continuous time martingale is considered.

It is tempting to argue in the following way. On $\{N_u < \infty\}$ the random variable $Y_{N_u} - u$ is the loss "above" ruin. The N_uth claim is larger than $u + X(S_{N_u}-)$ since it causes ruin. Further

$$Z_{N_u} = Y_{N_u} + X(S_{N_u}-) = (Y_{N_u} - u) + (u + X(S_{N_u}-))$$

and thus $Y_{N_u} - u$ ought to be exponentially distributed.

However, we do not consider this argument convincing, since the conditioning on $N_u < \infty$ has not been expressed in a way *exactly* corresponding to the conditioning in the definition of "lack of memory." Due to the independence in the risk model one might, however, accept the argument as a heuristic reasoning.

If we believe in the argument we have
$$E[e^{-R(u-Y_{N_u})} \mid N_u < \infty] = \int_0^\infty e^{Rz} \frac{1}{\mu} e^{-z/\mu} \, dz = \frac{1}{1-\mu R}$$
and thus we get
$$\Psi^0(u) = (1 - \mu R)e^{-Ru}. \tag{10}$$
Sparre Andersen (1957, p. 226) proved (10) in the special case where $\mu = 1$ and K^0 is a mixture of two exponential distributions.
□

Put
$$\mathcal{A}_1 = Y_{N_0} \text{ on } \{N_0 < \infty\} \qquad \text{and} \qquad A(y) = P\{\mathcal{A}_1 \leq y, \, N_0 < \infty\} \tag{11}$$
and note that $A(\infty) = \Psi^0(0)$. Thus A is a defective distribution. The defect $1 - A(\infty)$ is the probability that $\mathcal{A}_1$ is undefined. The curious notation $\mathcal{A}_1$ will soon get its explanation. By separating the cases $\mathcal{A}_1 > u$ and $\mathcal{A}_1 \leq u$ we have
$$\Psi^0(u) = A(\infty) - A(u) + \int_0^u \Psi^0(u-y) \, dA(y) \qquad \text{for } u \geq 0 \tag{12}$$
which, cf. (1.6), is a defective renewal equation. Assume, cf. (1.7), that there exists a constant κ such that
$$\int_0^\infty e^{\kappa y} \, dA(y) = 1. \tag{13}$$
Then, cf. (1.8),
$$e^{\kappa u}\Psi^0(u) = e^{\kappa u}(A(\infty) - A(u)) + \int_0^u e^{\kappa(u-y)}\Psi^0(u-y)e^{\kappa y} \, dA(y) \tag{14}$$
which is a proper renewal equation, and it then follows, cf. (1.9), that
$$\lim_{u \to \infty} e^{\kappa u}\Psi^0(u) = \frac{C_1}{C_2}, \tag{15}$$
where
$$C_1 = \int_0^\infty e^{\kappa y}(A(\infty) - A(y)) \, dy = \frac{1 - A(\infty)}{\kappa}$$
and
$$C_2 = \int_0^\infty y e^{\kappa y} \, dA(y)$$
provided κ, C_1, and C_2 exist in $(0, \infty)$ and that A is non-arithmetic, i.e., there exists no number d such that A is concentrated on $d, 2d, \ldots$. Formally, (15) looks like the Cramér-Lundberg approximation (**III**), but it is

– as it stands – almost useless, since the solution of (13) requires knowledge of A which is generally not known explicitly. From (8) and (15) it follows that $\kappa \geq R$ and our main result will be that $\kappa = R$.

Assume now that K^0 is continuous. This assumption is not necessary, but notation is much simplified. We shall now rely on the presentation of random walks given by Feller (1971, pp. 385 - 412). The reader is strongly recommended to consult Feller's presentation. The idea to use $\mathcal{A}_1$ in order to derive a renewal equation is due to Feller and (15) is formula (5.13) on p. 411 in Feller's book.

The reason for the "change of sign" made is to facilitate comparison with Feller. Further proceedings to simplify the comparison are a double numbering of the formulas, (15) would have been called (5.13) – (15), and the following "Translation of notation":

Translation of notation

Our notation	Feller's notation
$G(x)$	$F(x)$
γ	μ
Y_n	S_n
$\mathcal{A}_n$	$\mathcal{H}_n$
$A_n(y)$	$H_n(y)$
A_n	ψ_n
$\mathcal{D}_n$	$\mathcal{H}_n^-$
$D_n(y)$	$\rho_n(y)$

Put $T_1 = N_0$. $\mathcal{A}_1$ is called the first *ascending ladder* point. Define A_n by

$$A_n(y) = P\{T_1 = n, \, \mathcal{A}_1 \leq y\}$$
$$= P\{Y_1 \leq 0, \ldots, Y_{n-1} \leq 0, \, 0 < Y_n \leq y\} \qquad (1.4) - (16)$$

and note that $A(y) = \sum_{n=1}^{\infty} A_n(y)$. Since we have assumed K^0 to be continuous, we avoid the tedious distinction between strict and weak inequalities.

The section of the random walk following T_1 is a probabilistic replica of the whole random walk. The first ascending ladder point in that section is the *second* ascending ladder point in the whole random walk. Proceeding in that way we define an increasing sequence $\mathcal{A}_1, \mathcal{A}_2, \ldots$ of ladder points. Define the *renewal measure* $\mathbf{A}$ corresponding to A by

$$\mathbf{A}(y) = \sum_{n=0}^{\infty} A^{n*}(y), \qquad (1.8) - (17)$$

where A^{0*} is the atomic distribution with unit mass at the origin and A^{n*} is the nth convolution of A with itself. Obviously $\mathbf{A}(y) = 0$ for $y < 0$ while $\mathbf{A}(y)$ equals *one plus the expected number of ladder points* $\leq y$. Intuitively this means that we look at every ladder point $\mathcal{A}_n$ and count the number of them which are $\leq y$. Another way of doing this is to look at every variable Y_n and count the number of them which are ladder points $\leq y$. Y_n is a ladder point in $(0, y)$ if and only if

$$Y_n > Y_j \text{ for } j = 0, \ldots, n-1 \text{ and } Y_n \leq y. \quad (18)$$

Put $\mathbf{A}_0 = A^{0*}$ and let $\mathbf{A}_n(y)$ be the probability of the set given by (18). Then we have

$$\mathbf{A}(y) = \sum_{n=0}^{\infty} \mathbf{A}_n(y). \quad (3.2) - (19)$$

For fixed n we define n new variables by $X_1^* = X_n, \ldots, X_n^* = X_1$ and let $Y_0^*, Y_1^*, \ldots, Y_n^*$ be the corresponding random walk, i.e., $Y_0^* = 0$ and $Y_k^* = \sum_{j=1}^{k} X_j^*$. Obviously

$(Y_0^*, Y_1^*, \ldots, Y_n^*)$ has the same distribution as $(Y_0, Y_1, \ldots, Y_n)$.

Further, we have $Y_k^* = Y_n - Y_{n-k}$ and thus we have, cf. (18),

$$\{Y_n^* > Y_j^* \text{ for } j = 0, \ldots, n-1 \text{ and } Y_n^* \leq y\}$$
$$= \{Y_j > 0 \text{ for } j = 1, \ldots, n \text{ and } Y_n \leq y\}.$$

Thus we have

$$\mathbf{A}_n(y) = P\{Y_j > 0 \text{ for } j = 1, \ldots, n \text{ and } Y_n \leq y\} \quad (3.1) - (20)$$

for $n \geq 1$ and the following very useful lemma follows.

DUALITY LEMMA. *Feller (1971, p. 395). The renewal measure* $\mathbf{A}$ *admits of two interpretations. For every* $y > 0$ *the value* $\mathbf{A}(y)$ *equals*

(a)   *one plus the expected number of ladder points* $\leq y$; *and*

(b)   *one plus the expected number of events* $0 < Y_n < y$ *such that* $Y_k > 0$ *for* $k = 1, \ldots, n$.

The *descending* ladder points $\mathcal{D}_1, \mathcal{D}_2, \ldots$ are defined by symmetry, i.e., by changing $>$ into $<$. Then we have

$$D_n(y) = P\{Y_1 \geq 0, \ldots, Y_{n-1} \geq 0, Y_n < 0, Y_n \leq y\},$$

$$D(y) = \sum_{n=1}^{\infty} D_n(y),$$

$$\mathbf{D}_n(y) = P\{Y_n < Y_j \text{ for } j = 0, \ldots, n-1 \text{ and } Y_n \leq y\}$$
$$= P\{Y_j < 0 \text{ for } j = 1, \ldots, n \text{ and } Y_n \leq y\},$$

$$\mathbf{D}(y) = \sum_{n=0}^{\infty} D^{n*}(y) = \sum_{n=0}^{\infty} \mathbf{D}_n(y),$$

where $\mathbf{D}_0 = D^{0*} = \mathbf{A}_0$. Note that $\mathbf{D} - \mathbf{D}_0$ is concentrated on $(-\infty, 0)$. D is a proper distribution, i.e., $D(0) = 1$, and $E[\mathcal{D}_1]$ is finite, i.e., $E[\mathcal{D}_1] > -\infty$, when $-\infty < \gamma < 0$.

For $y \leq 0$ we have

$$D_{n+1}(y) = P\{Y_1 \geq 0, \ldots, Y_n \geq 0, Y_{n+1} \leq y\}$$

$$= \int_{0-}^{\infty} P\{Y_1 \geq 0, \ldots, Y_{n-1} \geq 0, Y_n \in (z, z+dz), Y_{n+1} \leq y\}$$

$$= \int_{0-}^{\infty} d\mathbf{A}_n(z)\, G(y-z) \qquad (3.5a) - (21)$$

and, in the same way, for $y > 0$,

$$\mathbf{A}_{n+1}(y) = P\{Y_1 \geq 0, \ldots, Y_n \geq 0,\ 0 < Y_{n+1} \leq y\}$$

$$= \int_{0-}^{\infty} P\{Y_1 \geq 0, \ldots, Y_{n-1} \geq 0, Y_n \in (z, z+dz),\ 0 < Y_{n+1} \leq y\}$$

$$= \int_{0-}^{\infty} d\mathbf{A}_n(z)\, (G(y-z) - G(-z)). \qquad (3.5b) - (22)$$

Summing over $n = 0, 1, \ldots$ yields

$$D(y) = \int_{0-}^{\infty} G(y-z)\, d\mathbf{A}(z) \qquad \text{for } y \leq 0 \qquad (3.7a) - (23)$$

and

$$\mathbf{A}(y) - 1 = \int_{0-}^{\infty} (G(y-z) - G(-z))\, d\mathbf{A}(z)$$

$$= \int_{0-}^{\infty} G(y-z)\, d\mathbf{A}(z) - D(0) \qquad \text{for } y > 0. \qquad (3.7b) - (24)$$

The convolution equations (23) and (24) *admit of exactly one probabilistically possible solution* $(D, \mathbf{A})$ (Feller 1971, pp. 401 - 402) where "probabilistically possible solution" means that D is a (possibly defective) distribution on $(-\infty, 0)$ and $\mathbf{A} - \mathbf{A}_0$ a measure on $(0, \infty)$ such that $\mathbf{A}(y) < \infty$ for $0 < y < \infty$.

Formulas analogous to (23) and (24) hold for $(A, \mathbf{D})$. We will explicitly use the analog of (23), where we have, for $y > 0$,

$$A_{n+1}(y) = P\{Y_1 \leq 0, \ldots, Y_n \leq 0,\ 0 < Y_{n+1} \leq y\}$$

$$= \int_{0-}^{\infty} P\{Y_1 \leq 0, \ldots, Y_{n-1} \leq 0, Y_n \in (z, z+dz),\ 0 < Y_{n+1} \leq y\}$$

$$= \int_{-\infty}^{0+} d\mathbf{D}_n(z)\, (G(y-z) - G(-z))$$

and, by summing,

$$A(y) = \int_{-\infty}^{0+} (G(y-z) - G(-z))\, d\mathbf{D}(z) \quad \text{for } y > 0. \tag{25}$$

Now we introduce the *associated* random walk induced by the random variables ${}^aX_1, {}^aX_2, \ldots$ with distribution ${}^aG$ given by

$$d\,{}^aG(x) = e^{Rx}\, dG(x),$$

where, of course, R is the Lundberg exponent. Since $g(r)$ is convex it follows from $g(0) = g(R)$ and $g'(0) < 0$ that $g'(R) > 0$. Thus

$$\,{}^a\gamma \stackrel{\text{def}}{=} E[{}^aX_k] = \int_{-\infty}^{\infty} xe^{Rx}\, dG(x) = g'(R) > 0$$

which implies that the corresponding ascending ladder points have a proper distribution ${}^aA$. It follows from Assumption 1.4 that ${}^a\gamma < \infty$ and thus also $E[{}^aA_1] < \infty$. If we write (23) and (24) in differential form, i.e.,

$$dD(y) = \int_{0-}^{\infty} G\{dy - z\}\, d\mathbf{A}(z) \quad \text{for } y \leq 0$$

and

$$d\mathbf{A}(y) = \int_{0-}^{\infty} G\{dy - z\} d\mathbf{A}(z) \quad \text{for } y > 0$$

and multiply with e^{Ry} it follows that

$$d\,{}^a\mathbf{D}(x) = e^{Rx}\, d\mathbf{D}(x) \quad \text{and} \quad d\,{}^a\mathbf{A}(x) = e^{Rx}\, d\mathbf{A}(x).$$

The same argument goes through for (A, D) and thus $d\,{}^aA(x) = e^{Rx}\, dA(x)$. Since ${}^aA$ is a proper distribution function we have, cf. (13),

$$\int_0^{\infty} e^{Ry}\, dA(y) = 1$$

and thus $\kappa = R$. Now the Cramér-Lundberg approximation

$$\lim_{u \to \infty} e^{Ru} \Psi^0(u) = \frac{1 - A(\infty)}{R \int_0^{\infty} ye^{Ry}\, dA(y)} \stackrel{\text{def}}{=} C^0 \tag{26}$$

follows from (15). (Note that (2) has not been explicitly used.) It follows immediately from our assumptions that $0 < C^0 < \infty$.

Since A is in general not known explicitly, the constant C^0 cannot be given a numerical value. Thorin (1974, p. 94) has given C^0 in a form which involves certain auxiliary functions. That form can probably be used for numerical purposes, although it is not at all as explicit as the constant in (III). This is not too surprising since the nice form of the constant in (III) does only hold when $F(0) = 0$, cf. Cramér (1955, p. 68).

It does, however, follow that R is the "right" exponent in (8). Because of the complexity of C^0 this is maybe the most important consequence of (26), since a Lundberg inequality with the "right" exponent is almost as informative as a Cramér-Lundberg approximation without a reasonably simple expression for the constant.

Let us now go back to (12). A formal solution is given by
$$\Psi^0(u) = 1 - [1 - A(\infty)]\mathbf{A}(u) \qquad (2.7) - (27)$$
since
$$A(\infty) - A(u) + \Psi^0 * A(u) = A(\infty) - A(u) + A(u) - [1 - A(\infty)][\mathbf{A}(u) - 1]$$
$$= A(\infty) + 1 - A(\infty) - [1 - A(\infty)]\mathbf{A}(u) = \Psi^0(u).$$

EXAMPLE 6. CONTINUED. (Feller 1971, pp. 405 and 410). Now $f(x) = \frac{1}{\mu}e^{-x/\mu}$ for $x \geq 0$, and thus
$$G'(x) = \int_{\max(0,-x/c)}^{\infty} \frac{1}{\mu} e^{-(x+cs)/\mu} \, dK^0(s)$$
which, for $x \geq 0$, reduces to
$$G'(x) = \frac{1}{\mu} e^{-x/\mu} \hat{k}^0(c/\mu)$$
or
$$G(x) = 1 - \hat{k}^0(c/\mu)\frac{1}{\mu} e^{-x/\mu}. \qquad (5.4) - (28)$$

From (25) we get
$$A(y) = \hat{k}^0(c/\mu) \int_{-\infty}^{0+} (e^{z/\mu} - e^{(z-y)/\mu}) \, d\mathbf{D}(z)$$
$$= (1 - e^{-y/\mu}) \, \hat{k}^0(c/\mu) \int_{-\infty}^{0+} e^{z/\mu} \, d\mathbf{D}(z). \qquad (29)$$

Letting $y \to \infty$ we get
$$A(y) = A(\infty)\left(1 - e^{-y/\mu}\right). \qquad (30)$$

It follows from (17) that, for $y \geq 0$,
$$\mathbf{A}(y) = 1 + \int_0^y \frac{A(\infty)}{\mu} \sum_{n=1}^{\infty} \left(\frac{A(\infty)}{\mu}\right)^{(n-1)} \frac{1}{(n-1)!} e^{-y/\mu} \, dy$$
$$= 1 + \int_0^y \frac{A(\infty)}{\mu} e^{-y(1-A(\infty))/\mu}$$
$$= \frac{1}{1 - A(\infty)} - \frac{A(\infty)}{1 - A(\infty)} e^{-y(1-A(\infty))/\mu} \qquad (31)$$

and thus, see (27),
$$\Psi^0(u) = A(\infty)e^{-u(1-A(\infty))/\mu}. \qquad (32)$$

From (26) we get $(1 - A(\infty))/\mu = R$ and thus (10) follows.
□

EXAMPLE 7. At the very end of Chapter 2 we mentioned that one practical application of the renewal model is when a classical risk model is inspected only at certain times. Let $\widetilde{S}_1, \widetilde{S}_2, \ldots$ denote the inspection epochs and assume that they form an ordinary renewal process with inter-occurrence time distribution $\widetilde{K}$. Assume that $\int_0^\infty e^{xs}\, d\widetilde{K}(s) < \infty$ for some $x > 0$. Let X be a classical risk process such that $\Psi(u) \sim Ce^{-Ru}$. Now we are interested in the ruin probability

$$\Psi_{\widetilde{K}}(u) = P\{\min_{n \geq 1} X(\widetilde{S}_n) < -u\}.$$

Put $Y_0 = 0$, $Y_n = -X(\widetilde{S}_n)$, and $X_n = Y_n - Y_{n-1}$ and note that $Y_0, Y_1, \ldots$ form a random walk. Obviously $X_1 = \sum_{k=1}^{N(\widetilde{S}_1)} Z_k - c\widetilde{S}_1$ and thus

$$g(r) \stackrel{\text{def}}{=} E[e^{rX_1}] = E[E\{e^{-rX(\widetilde{S}_1)} \mid \widetilde{S}_1\}] = E[e^{\widetilde{S}_1(\alpha h(r) - rc)}].$$

Since $g(R) = E[e^{\widetilde{S}_1 \cdot 0}] = 1$ it follows from (26) that

$$\Psi_{\widetilde{K}}(u) \sim C_{\widetilde{K}} e^{-Ru}$$

for some constant $C_{\widetilde{K}}$. Thus $\Psi_{\widetilde{K}}$ and Ψ differ asymptotically "only" in the constant and not in the Lundberg exponent.

Generally it seems difficult to relate $C_{\widetilde{K}}$ and C. The most interesting case is probably when $\widetilde{S}_1 = \Delta$ P-a.s., i.e., when $\widetilde{K}$ is a one-point distribution. The fact that $\widetilde{K}$ is not continuous is not important as long as F is non-arithmetic. In this case it follows from Cramér (1955, p. 75) that

$$C_{\widetilde{K}} \approx \frac{C}{\mu \rho R \cdot \alpha \Delta} \qquad \text{for large values of } \alpha \Delta.$$

$\alpha \Delta$ is the expected number of claims between inspections. The discussion in the appendix indicates that it is highly reasonable to consider large values of $\alpha \Delta$.
□

3.2 Stationary renewal models

Now we let N be a stationary renewal process. Then the distribution K of S_1 has density $k(s) = \alpha(1 - K^0(s))$ for $s \geq 0$ and its Laplace transform is

$$\hat{k}(v) = \frac{\alpha}{v}(1 - \hat{k}^0(v)). \tag{33}$$

Let $\Psi(u)$, without superscript, denote the ruin probability in the stationary case, while $\Psi^0(u)$ still denotes the ruin probability in the ordinary case. Put $\Phi(u) = 1 - \Psi(u)$ and $\Phi^0(u) = 1 - \Psi^0(u)$. By the "renewal" argument used in Section 1.1 we get the relation

$$\Phi(u) = \int_0^\infty k(s) \int_0^{u+cs} \Phi^0(u+cs-z)\, dF(z) ds. \qquad (34)$$

It is tempting to try to do something similar to what we did in Section 1.1 in order to derive (1.4). (In this section we do not rely on Feller, and consequently formula (1.4) means formula (4) in Section 1.) By changing the order of integration we get

$$\Phi(u) = \alpha \int_0^\infty (1 - K^0(s)) \int_0^{u+cs} \Phi^0(u+cs-z)\, dF(z)\, ds$$

$$= \alpha \int_0^\infty \int_s^\infty dK^0(v) \int_0^{u+cs} \Phi^0(u+cs-z)\, dF(z)\, ds$$

$$= \alpha \int_0^\infty dK^0(v) \int_0^v \int_0^{u+cs} \Phi^0(u+cs-z)\, dF(z)\, ds.$$

The change of variables $x = u + cs$ leads to

$$\Phi(u) = \frac{\alpha}{c} \int_0^\infty dK^0(v) \int_u^{u+cv} \int_0^x \Phi^0(x-z)\, dF(z)\, dx. \qquad (35)$$

Differentiation of both sides of (35), *provided* it is allowed, leads to

$$\Phi'(u)$$

$$= \frac{\alpha}{c} \int_0^\infty dK^0(v) \left\{ \int_0^{u+cv} \Phi^0(u+cv-z)\, dF(z) - \int_0^u \Phi^0(u-z)\, dF(z) \right\}$$

$$= \frac{\alpha}{c} \Phi^0(u) - \frac{\alpha}{c} \int_0^u \Phi^0(u-z)\, dF(z) \qquad (36)$$

which corresponds to (1.3). The last equality follows by the "renewal argument" applied to the ordinary case.

For future purpose the following simple result is given as a Lemma.

LEMMA 8. *Differentiation of (35) is allowed.*

PROOF: Put $\varphi(x) = \int_0^x \Phi^0(x-z)\, dF(z)$ and note that $0 \leq \varphi(x) \leq F(x) \leq 1$. Since

$$\left| \int_{u+\Delta}^{u+cv+\Delta} \varphi(x)\, dx - \int_u^{u+cv} \varphi(x)\, dx \right| \leq 2\Delta$$

the lemma follows by dominated convergence. ∎

Exactly as (1.4) and (1.5) follows from (1.3) we get

$$\Phi(u) = \Phi(0) + \frac{\alpha}{c} \int_0^u \Phi^0(u-z)(1-F(z))\, dz \qquad (37)$$

and

$$\Phi(\infty) = \Phi(0) + \frac{\alpha\mu}{c} \Phi^0(\infty). \qquad (38)$$

Since $\Phi(\infty) = \Phi^0(\infty) = 1$ when $c > \alpha\mu$ we have

$$\Psi(0) = \frac{\alpha\mu}{c} = \frac{1}{1+\rho} \qquad \text{when } c > \alpha\mu. \tag{39}$$

Thus **(I)** holds – *without any change* – also in this case, a result due to Thorin (1975, p. 97).

Exactly as (1.6) follows from (1.4) and **(I)**,

$$\Psi(u) = \frac{\alpha}{c}\int_u^\infty (1 - F(z))\, dz + \frac{\alpha}{c}\int_0^u \Psi^0(u-z)(1-F(z))\, dz \tag{40}$$

follows from (38) and (39).

EXAMPLE 6. CONTINUED. Consider again the case when $F(z) = 1 - e^{-z/\mu}$ and recall that $\Psi^0(u) = (1 - \mu R)e^{-Ru}$. From (40) we get

$$\Psi(u) = \frac{\alpha}{c}\int_u^\infty e^{-z/\mu}\, dz + \frac{\alpha}{c}\int_0^u (1-\mu R)e^{-R(u-z)}e^{-z/\mu}\, dz$$

$$= \frac{\alpha\mu}{c}e^{-u/\mu} + \frac{\alpha}{c}(1-\mu R)e^{-Ru}\int_0^u e^{-z(1-\mu R)/\mu}\, dz$$

$$= \frac{\alpha\mu}{c}e^{-u/\mu} + \frac{\alpha\mu}{c}e^{-Ru}\left(1 - e^{-u(1-\mu R)/\mu}\right) = \frac{\alpha\mu}{c}e^{-Ru}. \tag{41}$$

□

Now we consider Lundberg's inequality. Note that

$$\int_0^\infty e^{Rz}(1-F(z))\, dz = \frac{h(R)}{R}.$$

From (40) we get

$$\Psi(u) \leq \frac{\alpha}{c}\int_u^\infty (1-F(z))\, dz + \frac{\alpha}{c}\int_0^u e^{-R(u-z)}(1-F(z))\, dz$$

$$\leq \frac{\alpha}{c}\int_0^\infty e^{-R(u-z)}(1-F(z))\, dz = \frac{\alpha}{cR}h(R)e^{-Ru} \tag{42}$$

and thus Lundberg's inequality holds, with the difference that the constant may be larger than one. In the Poisson case we have $h(R) = cR/\alpha$ and (42) reduces to **(IV)**.

Finally we consider the Cramér-Lundberg approximation. By dominated convergence we get

$$\lim_{u\to\infty} e^{Ru}\Psi(u)$$

$$= \lim_{u\to\infty} \frac{\alpha}{c}e^{Ru}\int_u^\infty (1-F(z))\, dz +$$

$$\lim_{u\to\infty} \frac{\alpha}{c}\int_0^u e^{R(u-z)}\Psi^0(u-z)e^{Ru}(1-F(z))\, dz$$

$$= 0 + \frac{\alpha}{cR}h(R)\, C^0 \stackrel{\text{def}}{=} C, \tag{43}$$

a result due to Thorin (1975, p. 97). It follows immediately that $0 < C < \infty$.

3.3 Numerical illustrations

We shall now consider numerical illustrations of the R-values and of the ruin probabilities. We want especially to compare the Poisson case with the ordinary and the stationary renewal case. Let R_P and R denote the Lundberg exponent in the Poisson case and in the renewal case, respectively. Similarly $\Psi_P(u)$, $\Psi^0(u)$, and $\Psi(u)$ denote the ruin probability in the Poisson case, in the ordinary renewal case, and in the stationary renewal case. In all comparison α, c or ρ, and $F(z)$ will be the same. Recall that R_P and R are the positive solutions of

$$h(r) = \frac{cr}{\alpha} \quad \text{and} \quad (h(r)+1)\hat{k}^0(cr) = 1, \qquad (44)$$

respectively.

The value of the Lundberg exponent is a crude measure on the "dangerousness" of the risk business. With this crude measure the ordinary and the stationary renewal case are equally dangerous. If $R < R_P$ the renewal case is more "dangerous" than the Poisson case.

Consider a renewal model, and assume that $R < R_P$. It will follow from Theorem 4.22 that this is the case when N is a Cox process. From the definition of R_P it follows that $h(R_P) = cR_P/\alpha$ and thus $h(R) < cR/\alpha$ which implies that

$$\frac{\alpha}{cR} h(R) < 1 \quad \text{for} \quad R < R_P. \qquad (45)$$

Thus, in this case, **(IV)** holds and, see (43), $C < C^0$. This is rather natural, since in the stationary case we have

$$E[S_1] = \frac{E[S_1^0] + \text{Var}[S_1^0]}{2},$$

where S_1^0 is a random variable with distribution K^0 and large values of $\text{Var}[S_1^0]$ ought to correspond to "dangerous" cases. In the Cox case it can be shown (Kingmann 1964, p. 925) that $K(s) < K^0(s)$, which means that S_1 is "stochastically larger" than S_1^0. Intuitively this means that the company, in general, receives more premiums up to the first claim in the stationary case than in the ordinary case.

Let K_1^0 and K_2^0 be two inter-occurrence times distribution and let R_1 and R_2 be the corresponding Lundberg exponents. Then the following simple lemma holds.

LEMMA 9. *If $\hat{k}_1^0(v) \leq \hat{k}_2^0(v)$ for all $v > 0$, then $R_1 \geq R_2$.*

PROOF: We have
$$1 = (h(R_1) + 1)\hat{k}_1^0(cR_1) \leq (h(R_1) + 1)\hat{k}_2^0(cR_1)$$
and thus, since $(h(r) + 1)\hat{k}_1^0(cr)$ is convex, the lemma follows. ∎

Let K_D^0 be the one-point distribution
$$K_D^0(t) = \begin{cases} 0 & \text{if } t < 1/\alpha \\ 1 & \text{if } t \geq 1/\alpha \end{cases}$$
and let R_D be the corresponding Lundberg exponent. Then the very natural result follows.

COROLLARY 10. *Let R be the Lundberg exponent corresponding to any inter-occurrence times distribution K^0. Then $R \leq R_D$.*

PROOF: Let S_1^0 be a random variable with distribution K^0. It follows from Jensen's inequality that
$$\hat{k}^0(v) = E[e^{-vS_1^0}] \geq e^{-vE[S_1^0]} = e^{-v/\alpha} = \hat{k}_D^0(v)$$
and thus the corollary follows from Lemma 9. ∎

EXAMPLE 11.* Γ-DISTRIBUTED INTER-OCCURRENCE TIMES. As an example we consider the case when the inter-occurrence times density is given by
$$k^0(t) = \frac{t^{(1/\beta)-1}}{\beta^{1/\beta}\,\Gamma(1/\beta)} e^{-t/\beta} \qquad \text{for } t > 0$$
which is a Γ-distribution with $\alpha = 1$ and $\text{Var}[S_1^0] = \beta$. Let R_β denote the corresponding Lundberg exponent. Note that $R_1 = R_P$ and, formally, $R_0 = R_D$.

From example 2.42 it follows that the renewal process is

(i)   a Cox process if $\beta \geq 1$;

(ii)   a top process if $0 \leq \beta < 1$.

Since we only use this case as an illustration, we completely disregard the question of whether it is reasonable or not.

COROLLARY 12. *R_β is decreasing in β.*

PROOF: We have $\hat{k}_\beta^0(v) = (1 + \beta v)^{-1/\beta}$. Since the function $(1 + a/x)^x$ for $a > 0$ is increasing in x it follows that $\hat{k}_\beta^0(v)$, for fixed value of v, is increasing in β. Thus the corollary follows from Lemma 9. ∎

In order to get some quantitative apprehension of the dependence of R_β on β we consider large values of β. Recall that $\mu = E[Z_1]$ and $\sigma^2 = \text{Var}[Z_1]$.

---

* Since this example contains a remark, its end is marked by □ □.

3 Renewal models

PROPOSITION 13. *For large values of β we have*

$$R_\beta = \frac{r_0}{\beta} + \frac{\sigma^2 r_1}{\beta^2} + o\left(\frac{1}{\beta^2}\right), \tag{46}$$

where r_0 is the positive solution of

$$\mu r = \log(1 + cr) \tag{47}$$

and

$$r_1 = \frac{(1 + cr_0)r_0^2}{2(c - \mu - c\mu r_0)}.$$

PROOF: Put $r_0(\beta) = \beta R_\beta$. Thus $r_0(\beta)$ is the positive solution of

$$\left(h\left(\frac{r}{\beta}\right) + 1\right)(1 + cr)^{-1/\beta} = 1$$

or

$$\beta \log\left(1 + h\left(\frac{r}{\beta}\right)\right) - \log(1 + cr) = 0. \tag{48}$$

Put

$$f_\beta(r) = \beta \log(1 + h(r/\beta)) - \log(1 + cr) \quad \text{and} \quad f_\infty(r) = \mu r - \log(1 + cr).$$

The function $f_\infty(r)$ is continuous and convex on $[0, \infty)$. Since $f_\infty(0) = f_\infty(r_0) = 0$ and $f'_\infty(0) = \mu - c < 0$ it follows that $f'_\infty(r_0) > 0$. Further, $\lim_{\beta \to \infty} f_\beta(r) = f_\infty(r)$ and thus $\lim_{\beta \to \infty} r_0(\beta) = r_0$ which means that $r_0(\beta) = r_0 + o(1)$.

Put $r_1(\beta) = \beta \frac{r_0(\beta) - r_0}{\sigma^2}$. Then we have $r_0(\beta) = r_0 + \sigma^2 r_1(\beta)/\beta$ which satisfies

$$0 = \beta \log\left(1 + h\left(\frac{r_0}{\beta} + \frac{\sigma^2 r_1(\beta)}{\beta^2}\right)\right) - \log\left(1 + cr_0 + \frac{c\sigma^2 r_1(\beta)}{\beta}\right)$$

$$= \beta \log\left(1 + \frac{\mu r_0}{\beta} + \frac{\mu \sigma^2 r_1(\beta)}{\beta^2} + \frac{(\sigma^2 + \mu^2)r_0^2}{2\beta^2} + \frac{o(1)}{\beta^2}\right) -$$

$$\left[\log(1 + cr_0) + \log\left(1 + \frac{1}{\beta}\frac{c\sigma^2 r_1(\beta)}{1 + cr_0}\right)\right]$$

$$= \mu r_0 + \frac{\mu \sigma^2 r_1(\beta)}{\beta} + \frac{(\sigma^2 + \mu^2)r_0^2}{2\beta} - \frac{\mu^2 r_0^2}{2\beta} - \log(1 + cr_0) - \frac{1}{\beta}\frac{c\sigma^2 r_1(\beta)}{1 + cr_0} + \frac{o(1)}{\beta}$$

$$= \frac{\mu \sigma^2 r_1(\beta)}{\beta} + \frac{\sigma^2 r_0^2}{2\beta} - \frac{1}{\beta}\frac{c\sigma^2 r_1(\beta)}{1 + cr_0} + \frac{o(1)}{\beta}$$

or

$$\frac{cr_1(\beta)}{1 + cr_0} - \mu r_1(\beta) - \frac{r_0^2}{2} + o(1) = 0. \tag{49}$$

3.3 Numerical illustrations

Solving $r_1(\beta)$ in (49) we get

$$r_1(\beta) = \frac{(1+cr_0)r_0^2}{2(c-\mu-c\mu r_0)} + o(1) = r_1 + o(1).$$

Thus we get

$$R_\beta = \frac{r_0(\beta)}{\beta} = \frac{r_0 + \sigma^2 r_1(\beta)/\beta}{\beta} = \frac{r_0 + (\sigma^2 r_1 + o(1))/\beta}{\beta}$$

$$= \frac{r_0}{\beta} + \frac{\sigma^2 r_1}{\beta^2} + o\left(\frac{1}{\beta^2}\right). \quad \blacksquare$$

Now we also consider the case when the claims are Γ-distributed. For simplicity we put $\mu = 1$. Then R_β is the positive solution of

$$(1-\sigma^2 r)^{-1/\sigma^2}(1+c\beta r)^{-1/\beta} = 1$$

or

$$\frac{1}{\sigma^2}\log(1-\sigma^2 r) + \frac{1}{\beta}\log(1+c\beta r) = 0. \tag{50}$$

In the "extreme" case $\sigma^2 = 0$, which formally means that $Z_k = 1$ P-a.s., (50) reduces to

$$\beta r - \log(1+c\beta r) = 0$$

and thus, see (47), $R_\beta = r_0/\beta$. In the other "extreme" case $\beta = 0$ (50) reduces to

$$\log(1-\sigma^2 r) + \sigma^2 cr = 0$$

and thus $R_0 = \rho_0/\sigma^2$ where ρ_0 is the positive solution of $cr = -\log(1-r)$.

If σ^2 and β are of "the same order," we put $\sigma^2 = k\beta$. Then it follows from (50) that $R_\beta = \gamma_k/\beta$ where γ_k is the positive solution of

$$\log(1-kr) + k\log(1+cr) = 0.$$

For $c = 1.2$ we have

$$r_0 = 0.354199,$$
$$r_1 = -0.397223,$$
$$\rho_0 = 0.313698.$$

In Table 1 we give the values of R_β for some values of β and σ^2. For $\beta = 1$ this is one of the cases considered in Section 1.3 and for $\sigma^2 = 1$ there are exponentially distributed claims. In the last case we have

$$\frac{\Psi^0(u)}{\Psi(u)} = c(1-R_1)$$

and the table also illustrates the difference in the ordinary and the stationary case. For $c = 1.2$ the factor $c(1-R_1)$ ranges from 0.77 when $\beta \to 0$ to 1.2 when $\beta \to \infty$.

74   3 Renewal models

TABLE 1. Values of R_β for Γ-distributed inter-occurrence times and claims in the case $\alpha = \mu = 1$ and $c = 1.2$.

β	$\sigma^2 = 1$	$\sigma^2 = 10$	$\sigma^2 = 100$
0.001	0.3135	0.03130	0.003006
0.01	0.3110	0.03135	0.003114
0.1	0.2883	0.03110	0.003135
1	0.1667	0.02883	0.003110
10	0.03185	0.01667	0.002883
100	0.003503	0.003185	0.001667
1000	0.000354	0.000350	0.000318

Consider now the approximation $\frac{r_0}{\beta} + \frac{\sigma^2 r_1}{\beta^2}$ of R_β. If we again put $\sigma^2 = k\beta$ the approximation is reduced to $(r_0 + kr_1)/\beta$ and the comparison with Table 1 is simplified. Then we have

$$r_0 + kr_1 = \begin{cases} 0.3144 & \text{for } k = .1 \\ 0.3502 & \text{for } k = .01 \\ 0.3538 & \text{for } k = .001 \end{cases}.$$

REMARK 14. In the case $\beta = 1$ Grandell and Segerdahl (1971, pp. 151 - 153) considered a similar approximation of R_P for large values of σ^2. In that case

$$R_P = \frac{x_0}{\sigma^2} + \frac{x_1}{(\sigma^2)^2} + o\left(\frac{1}{(\sigma^2)^2}\right),$$

where x_0 is the positive solution of

$$cx + \log(1-x)$$

and

$$x_1 = \frac{c^2(1-x_0)x_0^2}{2[c(1-x_0) - 1]}.$$

For $c = 1.2$ we have

$$x_0 = 0.31370,$$
$$x_1 = -0.27560$$

and thus

$$\frac{x_0}{\sigma^2} + \frac{x_1}{(\sigma^2)^2} = \begin{cases} 0.02861 & \text{for } \sigma^2 = 10 \\ 0.003109 & \text{for } \sigma^2 = 100 \end{cases}.$$

EXAMPLE 15. MIXED EXPONENTIALLY DISTRIBUTED INTER-OCCURRENCE TIMES. Consider the inter-occurrence times distribution

$$K^0(t) = 1 - 0.25e^{-0.4t} - 0.75e^{-2t} \quad \text{for } t \geq 0,$$

where $\alpha = 1$ and $\text{Var}[S_1^0] = 2.5$.

The corresponding renewal process is a Cox process, see Example 2.37. Its intensity process $\lambda(t)$ is a two state Markov process with states 0 and 1.6 and with mean times in the states 2 and 10/3, respectively.

This case, originally considered by Sparre Andersen (1957, pp. 225 - 227), has been used by Thorin and Wikstad for numerical illustrations.

In Section 1.2 (1.35) we considered the claim distribution

$$F(z) = 1 - 0.0039793e^{-0.014631z} -$$
$$0.1078392e^{-0.190206z} - 0.8881815e^{-5.514588z} \quad \text{for } z \geq 0$$

which was used by Wikstad (1971).

In Tables 1.2 and 1.3 we have given values of R_P and $\Psi_P(u)$. There are only very small differences in the corresponding values of R and $\Psi(u)$. For $c = 1.1$, or $\rho = 10\%$, we get from Wikstad (1971, p. 152)

$$R = 0.0035,$$
$$\Psi^0(10) = 0.8125,$$
$$\Psi^0(100) = 0.5502,$$
$$\Psi^0(1000) = 0.0232.$$

Consider now the claim distribution

$$F(x) = 1 - 0.00001823254e^{-0.000496562z} -$$
$$0.0008425574e^{-0.00932298z} - 0.02405664e^{-0.09447445z} -$$
$$0.3114878e^{-0.7116063z} - 0.6635948e^{-3.675472z} \quad \text{for } z \geq 0$$

which was used by Thorin and Wikstad (1973) and Wikstad (1983).

TABLE 2. Ruin probabilities for mixed exponentially inter-occurrence times and claims in the case $c = 1.1$.

u	$\Psi_P(u)$	$\Psi^0(u)$	$\Psi(u)$
0	0.90909	0.93414	0.90909
100	0.47017	0.48126	0.47922
1000	0.20301	0.20414	0.20407
10000	0.00801	0.00807	0.00807

76   3 Renewal models

In Table 2 we give the values of ruin probabilities in the case $c = 1.1$. Those values are taken from Thorin and Wikstad (1973, p. 152) and Wikstad (1983, p. 47).

Also for this claim distribution there is very small difference in the ruin probabilities for the different models.
□

EXAMPLE 16. PARETO-DISTRIBUTED INTER-OCCURRENCE TIMES. Thorin and Wikstad (1973) considered the inter-occurrence time distribution

$$K^0(t) = 1 - (1 + 2t)^{-3/2} \qquad \text{for } t \geq 0,$$

where $\alpha = 1$ and $\text{Var}[S_1^0] = \infty$ when the claims are exponentially distributed with $\mu = 1$. In Table 3 we compare the Rs with the R_Ps. Since $R = 1 - \Psi^0(0)$ the R-values follow from Thorin and Wikstad (1973, p. 150).

TABLE 3.   Pareto-distributed inter-occurrence times and exponentially distributed claims.

c	R_P	R
1.05	0.04762	0.00146
1.10	0.09091	0.00540
1.15	0.13043	0.01124
1.20	0.16667	0.01855
1.25	0.20000	0.02699
1.30	0.23077	0.03628

□

Except for the Γ-distribution with $0 \leq \beta < 1$ all inter-occurrence time distributions discussed here are of the form

$$K^0(t) = \int_0^\infty (1 - e^{-t\theta}) \, dV(\theta) \qquad \text{for } t \geq 0, \qquad (51)$$

where V is a distribution function with $V(0) = 0$. These kind of distributions were discussed in Section 1.3 as claim distributions and in Section 2.2 as inter-occurrence time distributions. The corresponding renewal process is a Cox process, see Theorem 2.38. In the comparison of the Rs with the R_Ps we have $R \leq R_P$ when K^0 is of the form (51). That is no coincidence since, as mentioned, $R \leq R_P$ for Cox processes.

CHAPTER 4
Cox models

We shall now consider the case where the occurrence of the claims is described by a Cox process N. The first treatment of the ruin problem when the occurrence of the claims is described by a Cox process is due to Ammeter (1948). The special case he considered will be discussed Example 38. Reinhard (1984) derived an explicit formula for the ruin probability when the intensity process is a two-state Markov process and the claims are exponentially distributed. Asmussen (1989) considers the Cramér-Lundberg approximation when the intensity process is an M-state Markov process.

The main object of this chapter is to consider Lundberg inequalities, and that presentation is essentially based on Björk and Grandell (1988). In Section 4.1 we give some basic facts about Markov processes and consider special properties of the ruin probabilities when the intensity process is markovian. In Section 4.2 we begin the systematic treatment and present the basic idea, which is an extension of the "martingale approach" used in Section 1.1. In Section 4.3 we give a fairly complete analysis of the simple case when the intensity process has "independent jumps." In Section 4.4 we extend the analysis to the case of a Markov renewal intensity. In Section 4.5 we apply the results on the case of a markovian intensity. We further present a variation of the "martingale approach" which works in the Markov case. In Section 4.6 we compare, by simple examples, the Cox case with the Poisson case.

4.1 Markovian intensity: Preliminaries

Although we shall only be interested in rather special Markov processes, we shall start with a somewhat more general survey. A nice introduction to Markov processes, which the survey is highly based upon, is given by Feller (1971, pp. 321 - 357).

Basic Markov Process Theory

Let $Y = \{Y(t); t \geq 0\}$ be a stochastic process. The possible values S of $Y(t)$ is called the *state space*. Generally $S \subseteq \mathbf{R}$ but we shall also consider $S \subseteq \mathbf{R}^2$ later.

DEFINITION 1. Y is called a *Markov process* if
$$P\{Y(t) \in B \mid \mathcal{F}_s^Y\} = P\{Y(t) \in B \mid Y(s)\} \qquad P\text{-a.s.}$$
for all $t \geq s \geq 0$ and all $B \in \mathcal{B}(S)$.

Recall that $\mathcal{B}(S)$ is the Borel algebra on S.

The Poisson process, the Wiener process, the classical risk process, and the mixed Poisson process (see Definition 2.27) are all Markov processes.

DEFINITION 2. A (continuous) *transition probability* $Q_t(y, B)$, $t \geq 0$, $y \in S$, $B \in \mathcal{B}(S)$, is a function with the following properties:

(i) $Q_t(y, B)$ is a probability measure in B for fixed t and y;

(ii) $Q_t(y, B)$ is $\mathcal{B}(S)$-measurable in y for fixed t and B;

(iii) $Q_t(y, B)$ satisfies the Chapman-Kolmogorov equation
$$Q_{s+t}(y, B) = \int_S Q_t(y, dz) Q_s(z, B) \qquad \text{for } s,\, t \geq 0;$$

(iv) $Q_t(y, B) \to \begin{cases} 1 \text{ if } y \in B \\ 0 \text{ if } y \notin B \end{cases}$ as $t \to 0+$.

DEFINITION 3. A Markov process Y is said to be *homogeneous* or to have *stationary transition probabilities* if
$$P\{Y(t) \in B \mid Y(s) = y\} = Q_{t-s}(y, B) \qquad P\text{-a.s.}$$
for some transition probability $Q_t(y, B)$.

When nothing else is said it is understood that a Markov process is homogeneous. The non-homogeneous Poisson process and the mixed Poisson process are, however, simple examples of Markov processes which do not have stationary transition probabilities.

A homogeneous Markov process Y is determined by an *initial distribution* $p_0(B) \stackrel{\text{def}}{=} P\{Y(0) \in B\}$ and a transition probability $Q_t(y, B)$. Often we want Y to start in a given state. Then we put $p_0 = \delta_y$ where
$$\delta_y(B) = \begin{cases} 1 \text{ if } y \in B \\ 0 \text{ if } y \notin B \end{cases}.$$

DEFINITION 4. If for some functions v and $w : S \to \mathbf{R}$
$$\int_S \frac{v(z) - v(y)}{\Delta} Q_\Delta(y, dz) \to w(y) \qquad \text{as } \Delta \to 0+ \qquad (1)$$

(in the sense of uniform convergence) we put $w = Av$. The operator so defined is called the *generator* or the *infinitesimal operator* of Y. The function v is said to be in the domain of A.

The domain of a generator contains "enough" functions to ensure that the generator determines the transition probabilities uniquely, see Feller (1971, p. 456).

When it is possible to use all $y \in S$ as initial states, (1) may be replaced with

$$\frac{1}{\Delta} E[v(Y(\Delta)) - v(y) \mid Y(0) = y] \to w(y) \qquad \text{as } \Delta \to 0+. \qquad (2)$$

The following well-known result gives a connection between Markov processes and martingales.

PROPOSITION 5. (Dynkin's theorem.) *Let Y be a Markov process with generator A and v a function in the domain of A such that* $Av \equiv 0$. Then M, defined by $M(t) = v(Y(t))$, is an $\mathbf{F}^Y$-martingale.*

PROOF: We shall show that

$$E[M(t+s) - M(s) \mid \mathcal{F}_s^Y] = E[M(t+s) \mid \mathcal{F}_s^Y] - M(s) = 0.$$

Choose any $\epsilon > 0$. From the uniform convergence in (1) it follows that there exists $\delta > 0$ such that

$$\left| \int_S (v(z) - v(y)) Q_\Delta(y, dz) \right| < \epsilon \Delta \qquad \text{for all } \Delta < \delta \text{ and all } y \in S. \qquad (3)$$

Choose n and $\Delta < \delta$ such that $t/n = \Delta$. Then we have

$$\left| E[M(t+s) - M(s) \mid \mathcal{F}_s^Y] \right|$$

$$= \left| E\left[\sum_{k=1}^{n} M(s + k\Delta) - M(s + (k-1)\Delta) \mid \mathcal{F}_s^Y \right] \right|$$

$$\leq \sum_{k=1}^{n} \left| E[M(s + k\Delta) - M(s + (k-1)\Delta) \mid \mathcal{F}_s^Y] \right|$$

$$= \sum_{k=1}^{n} \left| E\big[E[M(s + k\Delta) - M(s + (k-1)\Delta) \mid \mathcal{F}_{s+(k-1)\Delta}^Y] \mid \mathcal{F}_s^Y \big] \right|$$

$$= \sum_{k=1}^{n} \left| E\big[E[M(s + k\Delta) - M(s + (k-1)\Delta) \mid Y(s + (k-1))] \mid \mathcal{F}_s^Y \big] \right|,$$

---

* Strictly speaking we must also require $E[|v(Y(0))|] < \infty$, which is a condition both on v and p_0. This is automatically fulfilled if v is bounded or if $p_0 = \delta_y$. In our applications this condition causes no problem.

where the last equality follows since Y is markovian. It follows from (3) that

$$\left| E[M(s+k\Delta) - M(s+(k-1)\Delta) \mid Y(s+(k-1))] \right|$$
$$= \left| E\bigl[v(Y(s+k\Delta)) - v(Y(s+(k-1)\Delta)) \mid Y(s+(k-1))\bigr] \right| < \epsilon\Delta$$

and thus

$$\left| E[M(t+s) - M(s) \mid \mathcal{F}_s^Y] \right| < \epsilon n \Delta = \epsilon t$$

which, since ϵ was arbitrarily chosen, the proposition is proved. ∎

DEFINITION 6. A Markov process Y is called a (markovian) *jump process* if it has piecewise constant realizations.

Let Y be a jump process. Given that $Y(t) = y$, the waiting time to the next jump is exponentially distributed with – say – mean $1/\eta(y)$. The probability that the following jump leads to a point in B is denoted by $p_L(y, B)$. $\eta(y)$ is called the *intensity function* and $p_L(y, B)$ the *jump measure*. (The subscript L will get its explanation later.) We shall, essentially, only consider jump processes. The Wiener process is a diffusion process and not a jump process.

The following lemma will be used later.

LEMMA 7. *The generator H of a jump process is given by*

$$(Hv)(y) = \eta(y) \int_S v(z) p_L(y, dz) - \eta(y)v(y). \tag{4}$$

PROOF: We have

$$E[v(Y(\Delta)) \mid Y(0) = y] = (1 - \eta(y)\Delta)v(y) + \eta(y)\Delta \int_S p_L(y, dz)v(z) + o(\Delta)$$

and the lemma follows from (2). ∎

Consider the classical risk process X given by (1.1). Formally X is not a jump process.

LEMMA 8. *The generator G_α of the classical risk process is given by*

$$(G_\alpha v)(y) = cv'(y) + \alpha \int_0^\infty v(y-z) \, dF(z) - \alpha v(y). \tag{5}$$

PROOF: Since $X(0) = 0$ by definition we consider $Y(t) = y + X(t)$ which has the same generator. We have

$$E[v(Y(\Delta)) \mid Y(0) = y]$$
$$= (1 - \alpha\Delta)v(y + c\Delta) + \alpha\Delta \int_0^\infty v(y + c\Delta - z) \, dF(z) + o(\Delta)$$
$$= v(y) + c\Delta v'(y) - \alpha\Delta v(y) + \alpha\Delta \int_0^\infty v(y-z) \, dF(z) + o(\Delta)$$

and the lemma follows from (2). ∎

REMARK 9. Lemma 8 also follows from Lemma 7, applied to $Y(t) = y + X(t) - ct$, which is a jump process with $\eta(y) \equiv \alpha$ and $p_L(y, dz) = F(y - dz)$.
□

In connection with Markov processes it is usual to derive differential equations either by a backward or a forward argument. Consider the time interval $(0, t + \Delta]$. In a backward argument the interval is divided into a "short" interval $(0, \Delta]$ and a "long" interval $(\Delta, t + \Delta]$ and in a forward argument the interval is divided into a "long" interval $(0, t)$ and a "short" interval $(t, t + \Delta]$. The idea is to consider the possible changes in the short interval. Generally the backward argument goes easily through while the forward argument requires additional assumptions. We may note that the "differential argument" used to derive (1.3) was, in fact, a backward argument.

Consider now, in a jump process, the one-dimensional distribution

$$p_t(B) \stackrel{\text{def}}{=} P\{Y(t) \in B\}.$$

A forward argument *formally* yields

$$p_{t+\Delta}(B) = \int_S p_t(dy) Q_\Delta(y, B)$$

$$= \int_B p_t(dy)(1 - \eta(y)\Delta + o(\Delta)) + \int_S p_t(dy)(\eta(y)\Delta p_L(y, B) + o(\Delta)) + o(\Delta)$$

$$= p_t(B) + \Delta \left(-\int_B p_t(dy)\eta(y) + \int_S p_t(dy)\eta(y)p_L(y, B) \right) + o(\Delta) \quad (6)$$

and thus

$$\frac{\partial p_t(B)}{\partial t} = -\int_B p_t(dy)\eta(y) + \int_S p_t(dy)\eta(y)p_L(y, B). \quad (7)$$

The problem in this derivation is that the $o(\Delta)$s appearing under the integrals in (6) depend on y. If, for example, $\eta(y)$ is bounded in y there is no problems and (7) holds. For details we refer to Feller (1971, pp. 327 - 328).

If (7) really holds and if there exists an initial distribution p_0 such that $\frac{\partial p_t(B)}{\partial t} \equiv 0$ then the jump process Y becomes stationary. This leads to the following definition.

DEFINITION 10. Let Y be a (markovian) jump process. A probability measure q_L on $(S, \mathcal{B}(S))$ such that

$$\int_B q_L(dy)\eta(y) = \int_S q_L(dy)\eta(y)p_L(y, B) \quad (8)$$

is called a *stationary initial* distribution of Y.

We shall only consider jump processes – also in non-markovian cases – as models for the intensity. In the markovian case we shall be interested in two sub-classes of jump processes, namely jump processes with "independent jumps" and M-state Markov processes.

DEFINITION 11. A (markovian) jump process is called a Markov process with *independent jumps* if its jump measure is independent of y, i.e., if
$$p_L(y, B) = p_L(B).$$
In this case (8) is reduced to
$$\int_B q_L(dy)\eta(y) = p_L(B) \int_S q_L(dy)\eta(y). \tag{9}$$
From (9) and $q_L(S) = 1$ it follows that
$$q_L(dy) = \frac{p_L(dy)}{\eta(y)} \bigg/ \int_S \frac{p_L(dz)}{\eta(z)}, \tag{10}$$
where $\int_S \frac{p_L(dz)}{\eta(z)}$ can be interpreted as the mean duration between two successive jumps. It will later be seen that a unique initial distribution which makes Y stationary exists as soon as $\int_S \frac{p_L(dz)}{\eta(z)} < \infty$ whether $\eta(y)$ is bounded or not.

DEFINITION 12. A (markovian) jump process is called an M-state Markov process if its state space S is an M-point space $\{\alpha_1, \alpha_2, \ldots, \alpha_M\}$.

In this case functions, measures, and operators are reduced to vectors and matrices. Let
$$\mathbf{A} = (a_{ij}) \text{ be the matrix } \begin{pmatrix} a_{11} & \cdots & a_{1M} \\ \vdots & & \vdots \\ a_{M1} & \cdots & a_{MM} \end{pmatrix} \text{ and}$$
$$\mathbf{a} = (a_i) \text{ the vector } \begin{pmatrix} a_1 \\ \vdots \\ a_M \end{pmatrix}$$
We shall use the following notation:

$\mathbf{a}^T = (a_1, \ldots, a_M)$;

$(\mathbf{a})_i = a_i$ and $(\mathbf{A})_{ij} = a_{ij}$;

$(\mathbf{A})_{i\cdot} = (a_{i1}, \ldots, a_{M1})$ and $(\mathbf{A})^T_{\cdot j} = (a_{1j}, \ldots, a_{Mj})$;

$\mathbf{I} = (\delta_{ij})$ is the identity matrix;

$\mathbf{1}^T = (1)^T$ is the vector $(1, \ldots, 1)$;

$\mathbf{0}^T = (0)^T$ is the vector $(0, \ldots, 0)$;

$d(\mathbf{A})$ is the diagonal matrix $(\delta_{ij} a_{ij})$;

$d(\mathbf{a})$ is the diagonal matrix $(\delta_{ij} a_i)$.

A vector **a** is called a *distribution* if $a_i \geq 0$ and $\mathbf{a}^T\mathbf{1} = 1$. **A** is called a *stochastic matrix* if $(\mathbf{A})_{i\cdot}$, $i = 1, \ldots, M$, is a distribution.

Let Y be an M-state Markov process with state space $\{\alpha_1, \alpha_2, \ldots, \alpha_M\}$. Put
$$\eta_i = \eta(\alpha_i), \qquad \boldsymbol{\eta} = (\eta_i), \quad \text{and} \quad p_{ij} = p_L(\alpha_i, \{\alpha_j\}).$$
Then $\mathbf{P} = (p_{ij})$ is a stochastic matrix. Assume that all states communicate, i.e., that for all i and j there exist $i_1, i_2, \ldots, i_n$ such that
$$p_{ii_1} p_{i_1 i_2} \cdots p_{i_{n-1} i_n} p_{i_n i} > 0$$
and that $p_{ii} = 0$ for all i. Although the reader is certainly familiar with M-state Markov processes we shall derive some well-known properties from the general results. The generator $\mathbf{H} \stackrel{\text{def}}{=} (\eta_{ij})$ is in this case generally called the *intensity matrix*. It follows from (4) that
$$(\mathbf{Hv})_i = \eta_i (\mathbf{Pv})_i - \eta_i v_i = (d(\boldsymbol{\eta})\mathbf{Pv})_i - (d(\boldsymbol{\eta})\mathbf{v})_i,$$
where $\mathbf{v} = (v_i)$, and thus
$$\mathbf{H} = d(\boldsymbol{\eta})(\mathbf{P} - \mathbf{I}) \quad \text{or} \quad \eta_{ij} = \begin{cases} \eta_i p_{ij} & \text{if } i \neq j \\ -\eta_i & \text{if } i = j \end{cases}. \tag{11}$$

A distribution $\mathbf{q} = (q_i)$ is a stationary initial distribution, see (8), if
$$q_i \eta_i = \sum_{j=1}^{M} q_j \eta_j p_{ji} = (\mathbf{q}^T d(\boldsymbol{\eta})\mathbf{P})_i \quad \text{or} \quad \mathbf{q}^T d(\boldsymbol{\eta}) = \mathbf{q}^T d(\boldsymbol{\eta})\mathbf{P}, \quad \text{i.e.,}$$
$$\mathbf{q}^T \mathbf{H} = \mathbf{0}^T. \tag{12}$$

For $M = 2$ we have $\mathbf{P} = \begin{pmatrix} 0 & 1 \\ 1 & 0 \end{pmatrix}$ and $\mathbf{H} = \begin{pmatrix} -\eta_1 & \eta_1 \\ \eta_2 & -\eta_2 \end{pmatrix}$ and thus
$$\left. \begin{array}{l} q_1 = \eta_2/(\eta_1 + \eta_2) \\ q_2 = \eta_1/(\eta_1 + \eta_2) \end{array} \right\}. \tag{13}$$

□ □

Let $\lambda(t)$ be a (markovian) jump process with $S \subset \mathbf{R}_+$, intensity function $\eta(x)$, jump measure $p_L(x, B)$, and initial distribution p_L. Assume that there exists a stationary initial distribution q_L with mean α. The claims are described by a Cox process N with $\lambda(t)$ as its intensity process. Assume that $\rho > 0$, where the relative safety loading, exactly as in the Poisson and the renewal cases, is defined by
$$\rho = \frac{c - \alpha\mu}{\alpha\mu}.$$
Let $\Psi_x(u)$ denote the ruin probability when $\lambda(0) = x$, i.e., when $p_0 = \delta_x$, and let $\Psi(u)$ denote the ruin probability in the stationary case, i.e., when

$p_0 = q_L$. Put
$$\Phi_x(u) = 1 - \Psi_x(u) \quad \text{and} \quad \Phi(u) = 1 - \Psi(u).$$
By a "backward differential argument" we get, comparing the derivation of (1.3),
$$\Phi_x(u) = (1 - x\Delta - \eta(x)\Delta)\Phi_x(u + c\Delta) + x\Delta \int_0^{u+c\Delta} \Phi_x(u + c\Delta - z)\, dF(z) +$$
$$\eta(x)\Delta \int_S p_L(x, dy)\Phi_y(u + c\Delta) + o(\Delta)$$
and thus
$$c\Phi'_x(u) = x\Phi_x(u) + \eta(x)\Phi_x(u) - x \int_0^u \Phi_x(u - z)\, dF(z) -$$
$$\eta(x) \int_S p_L(x, dy)\Phi_y(u) \quad \text{for } x \in S \tag{14}$$

which is essentially formula (5.34) in Reinhard (1984, p. 38).

Integration of (14) over $(0, t)$ yields, comparing the derivation of (1.4),
$$c[\Phi_x(t) - \Phi_x(0)]$$
$$= x \int_0^t \Phi_x(t - z)(1 - F(z))\, dz +$$
$$\eta(x) \int_0^t \left[\Phi_x(u) - \int_S p_L(x, dy)\Phi_y(u)\right] du \tag{15}$$
which is (5.36) in Reinhard (1984, p. 38). Since
$$\int_S q_L(dx)\eta(x)p_L(x, dy) = q_L(dy)\eta(y),$$
integration with respect to q_L reduces (15) to
$$c[\Phi(t) - \Phi(0)] = \int_0^t \left[\int_S x\Phi_x(t - z)\, q_L(dx)\right](1 - F(z))\, dz. \tag{16}$$
By monotone convergence it follows from (16), as $t \to \infty$, that
$$c[\Phi(\infty) - \Phi(0)] = \mu \int_S x\Phi_x(\infty)\, q_L(dx) \tag{17}$$
and thus
$$\Psi(0) = \frac{\alpha\mu}{c} + \Psi(\infty) - \frac{\mu}{c} \int_S x\Psi_x(\infty)\, q_L(dx). \tag{18}$$
In "non-pathological" cases $\Psi_x(\infty) = 0$ q_L-a.s. and then (18) reduces to
$$\Psi(0) = \frac{\alpha\mu}{c} = \frac{1}{1 + \rho} \quad \text{when } c > \alpha\mu. \tag{19}$$
Thus we have again found a class of stationary risk models where (**I**) holds.

REMARK 13. There do exist "pathological" cases where (19) does not hold. If, for example, $p_L(x, B) = \delta_x(B)$ it follows that $\lambda(t) = \lambda(0)$ for all t, i.e., N is a mixed Poisson process. Then any distribution U on $\mathbf{S}$ works as stationary initial distribution. Since

$$\Psi_x(\infty) = \begin{cases} 0 \text{ if } x < c/\mu \\ 1 \text{ if } x \geq c/\mu \end{cases}$$

it follows that

$$\Psi(0) = \frac{\mu}{c} \int_0^{c/\mu} x \, dU(x) + \left(1 - U\left(\frac{c}{\mu}\right)\right). \tag{20}$$

In Theorem 5.6 we shall show that (20) is the general version of (I) and, with a proper interpretation of U, holds for all stationary risk models. □

A sufficient condition – see the proof of Theorem 5.6 – for

$$\Psi_x(\infty) = 0 \quad q_L\text{-a.s. is } \lim_{t \to \infty} N(t)/t = \alpha \quad P\text{-a.s.}$$

It especially follows that $\Psi_x(\infty) = 0$ q_L-a.s. if $\lambda(t)$ is an M-state Markov process, since all states are assumed to communicate.

Now we consider the case where the claims are exponentially distributed with mean μ and where $\lambda(t)$ is a two-state Markov process with state space $\{\alpha_1, \alpha_2\}$ where, and this is no restriction, $\alpha_1 \leq \alpha_2$. Put

$$\Psi_k(u) = \Psi_{\alpha_k}(u) \quad \text{and} \quad \Phi_k(u) = \Phi_{\alpha_k}(u)$$

and note that

$$\Phi_k(\infty) = 1. \tag{21}$$

In this case (14) is reduced to

$$\left. \begin{aligned} c\Phi_1'(u) &= (\alpha_1 + \eta_1)\Phi_1(u) - \frac{\alpha_1}{\mu}\int_0^u \Phi_1(z)e^{-(u-z)/\mu}\,dz - \eta_1\Phi_2(u) \\ c\Phi_2'(u) &= (\alpha_2 + \eta_2)\Phi_2(u) - \frac{\alpha_2}{\mu}\int_0^u \Phi_2(z)e^{-(u-z)/\mu}\,dz - \eta_2\Phi_1(u) \end{aligned} \right\} \tag{22}$$

from which, of course, it follows that

$$\left. \begin{aligned} c\Phi_1'(0) &= (\alpha_1 + \eta_1)\Phi_1(0) - \eta_1\Phi_2(0) \\ c\Phi_2'(0) &= (\alpha_2 + \eta_2)\Phi_2(0) - \eta_2\Phi_1(0) \end{aligned} \right\}. \tag{23}$$

Differentiation of (22) leads to, compare Example 1.7,

$$\left. \begin{aligned} c\Phi_1''(u) &= \quad (\alpha_1 + \eta_1 - \frac{c}{\mu})\Phi_1'(u) - \eta_1\Phi_2'(u) + \frac{\eta_1}{\mu}\Phi_1(u) - \frac{\eta_1}{\mu}\Phi_2(u) \\ c\Phi_2''(u) &= -\eta_2\Phi_1'(u) + (\alpha_2 + \eta_2 - \frac{c}{\mu})\Phi_2'(u) - \frac{\eta_2}{\mu}\Phi_1(u) + \frac{\eta_2}{\mu}\Phi_2(u) \end{aligned} \right\} \tag{24}$$

with boundary conditions (21) and (23). The differential system (24) is (6.10) in Reinhard (1984, p. 40).

REMARK 14. Before discussing (24) we observe that N is a Poisson process when $\alpha_1 = \alpha_2$ and, see Example 2.37, a renewal process when $\alpha_1 = 0$. In both cases we know that (II) holds, i.e.,

$$\Psi(u) = \frac{\alpha\mu}{c} e^{-Ru}.$$

In the Poisson case

$$R = \frac{\rho}{\mu(1+\rho)} \quad \left(= \frac{1}{\mu} - \frac{\alpha}{c} \right).$$

In the renewal case, see (3.7) and (2.17), R is the smallest positive solution of

$$\frac{\alpha_2\eta_1 + c\alpha_2 r}{\alpha_2\eta_1 + cr(\alpha_2 - \eta_1 + \eta_2) + c^2 r^2} \cdot \frac{1}{1-\mu r} = 1. \tag{25}$$

Note that (25) has probabilistic meaning only for $0 \leq r < 1/\mu$. By routine calculations it follows that (25) is equivalent to

$$r\beta(r) \stackrel{\text{def}}{=} r\left(r^2 + r\left(\frac{\alpha_2 + \eta_1 + \eta_2}{c} - \frac{1}{\mu}\right) - \left(\frac{\eta_1 + \eta_2}{c\mu} - \frac{\alpha_2\eta_1}{c^2}\right) \right) = 0. \tag{26}$$

Since

$$0 < \frac{c - q_2\alpha_2\mu}{q_2\alpha_2\mu} = \frac{c(\eta_1 + \eta_2) - \eta_1\alpha_2\mu}{\alpha_2\eta_1\mu} = \frac{c^2}{\alpha_2\eta_1}\left(\frac{\eta_1 + \eta_2}{c\mu} - \frac{\alpha_2\eta_1}{c^2}\right)$$

it follows that the solutions of $\beta(r) = 0$ have opposite signs and thus

$$R = -\frac{1}{2}\left(\frac{\alpha_2 + \eta_1 + \eta_2}{c} - \frac{1}{\mu}\right) +$$

$$\sqrt{\frac{1}{4}\left(\frac{\alpha_2 + \eta_1 + \eta_2}{c} - \frac{1}{\mu}\right)^2 + \frac{\eta_1 + \eta_2}{c\mu} - \frac{\alpha_2\eta_1}{c^2}}. \tag{27}$$

□

Assume now that $\alpha_1 < \alpha_2$. Reinhard (1984, p. 40) carried through a detailed analysis of (24) and therefore we omit details.

Putting

$$\Phi_1(u) = ae^{-ru} \quad \text{and} \quad \Phi_2(u) = ad(r)e^{-ru}$$

in (24) we get

$$\left. \begin{array}{l} c\,r^2 = -(\alpha_1 + \eta_1 - \dfrac{c}{\mu})r + \eta_1 d(r)\,r + \dfrac{\eta_1}{\mu} - \dfrac{\eta_1}{\mu}d(r) \\[2mm] cd(r)\,r^2 = \eta_2 r - (\alpha_2 + \eta_2 - \dfrac{c}{\mu})d(r)\,r - \dfrac{\eta_2}{\mu} + \dfrac{\eta_2}{\mu}d(r) \end{array} \right\}. \tag{28}$$

4.1 Markovian intensity: Preliminaries

Solving $d(r)$ in one of the equations (28) and putting the solution in the other, we get after routine calculations

$$d(r) = \frac{c\mu r^2 + (\alpha_1\mu + \eta_1\mu - c)r - \eta_1}{\eta_1\mu r - \eta_1} \tag{29}$$

or

$$d(r) = \frac{\eta_2\mu r - \eta_2}{c\mu r^2 + (\alpha_2\mu + \eta_2\mu - c)r - \eta_2}$$

and

$$r\gamma(r) \stackrel{\text{def}}{=} r\left(r^3 + r^2\left(\frac{\alpha_1 + \alpha_2 + \eta_1 + \eta_2}{c} - \frac{2}{\mu}\right) + r\left(\frac{\alpha_1\alpha_2 + \alpha_1\eta_2 + \alpha_2\eta_1}{c^2} - \frac{\alpha_1 + \alpha_2 + 2\eta_1 + 2\eta_2}{c\mu} + \frac{1}{\mu^2}\right) - \left(\frac{\alpha_1\eta_2 + \alpha_2\eta_1}{c^2\mu} - \frac{\eta_1 + \eta_2}{c\mu^2}\right)\right) = 0. \tag{30}$$

REMARK 15. For $r = 1/\mu$ we have, see (29),

$$d(1/\mu) = \frac{\alpha_1}{0} = \frac{0}{\alpha_2}.$$

Since $\gamma(1/\mu) = \frac{\alpha_1\alpha_2}{c^2\mu}$ this happens only when $\alpha_1 = 0$, i.e., (see Remark 14) only in the renewal case.
□

If (30) has distinct solutions $r_1, \ldots, r_4$ it follows from the theory of systems of differential equations that

$$\Phi_1(u) = \sum_{k=1}^{4} a_k e^{-r_k u} \quad \text{and} \quad \Phi_2(u) = \sum_{k=1}^{4} a_k d(r_k) e^{-r_k u}, \tag{31}$$

where $a_1, \ldots, a_4$ are determined by (21) and (23).
Put

$$R_1 = \frac{1}{\mu} - \frac{\alpha_1}{c} \quad \text{and} \quad R_2 = \frac{1}{\mu} - \frac{\alpha_2}{c}$$

and note that $R_1 > R_2$. It follows from Reinhard (1984, p. 41) that

$$r_1 < 0 = r_2 \leq \max(0, R_2) < r_3 < R_1 < r_4. \tag{32}$$

Thus the solutions are distinct. It follows from (21) that $a_1 = 0$ and $a_2 = 1$. Note that $d(0) = 1$. From (23) we get

$$\left.\begin{aligned}
[\eta_1 d(r_3) - (cr_3 + \eta_1 + \alpha_1)]a_3 + [\eta_1 d(r_4) - (cr_4 + \eta_1 + \alpha_1)]a_4 &= \alpha_1 \\
[\frac{\eta_2}{d(r_3)} - (cr_3 + \eta_2 + \alpha_2)]d(r_3)\, a_3 + [\frac{\eta_2}{d(r_4)} - (cr_4 + \eta_2 + \alpha_2)]d(r_4)\, a_4 &= \alpha_2
\end{aligned}\right\}$$

which, in view of (29), is equivalent to

$$\left.\begin{aligned}
\frac{1}{\mu r_3 - 1} a_3 + \frac{1}{\mu r_4 - 1} a_4 &= 1 \\
\frac{d(r_3)}{\mu r_3 - 1} a_3 + \frac{d(r_4)}{\mu r_4 - 1} a_4 &= 1
\end{aligned}\right\} \tag{33}$$

and thus
$$-a_3 = \frac{(1 - \mu r_3)(d(r_4) - 1)}{d(r_3) - d(r_4)}. \tag{34}$$

REMARK 16. It is natural to interpret α_1 as the "normal" risk which now and then is replaced by the "increased" risk α_2. In automobile insurance α_1 might be the risk under "normal" conditions while α_2 might be the risk under "bad" – think of slippery roads, foggy days, high traffic volume, and so on – conditions. Certainly we do not claim that a two-state Markov process is a realistic description of risk fluctuation in this case. Anyhow, it may be natural to let the risks α_1 and α_2 affect not only the frequency of the claims but also their sizes.

Reinhard (1984), in fact, took this into account and allowed for different μs, and also different cs, in the two states.
□

Primarily we are interested in $\Psi(u) = q_1\Psi_1(u) + q_2\Psi_2(u)$ and not in Φ_1 and Φ_2. Put $R = r_3$ and $\widetilde{R} = r_4$. Then we have, see (13) and (19),
$$\Psi(u) = Ce^{-Ru} + \widetilde{C}e^{-\widetilde{R}u}, \tag{35}$$
where
$$C = -\frac{\eta_2 a_3 + \eta_1 d(R)a_3}{\eta_1 + \eta_2} = \frac{(1 - \mu R)(1 - d(\widetilde{R}))}{d(R) - d(\widetilde{R})} \frac{\eta_2 + \eta_1 d(R)}{\eta_1 + \eta_2}$$
and
$$\widetilde{C} = \frac{\alpha\mu}{c} - C$$
so (II) does not hold unless $C = \alpha\mu/c$. We note that
$$\lim_{u \to \infty} e^{Ru}\Psi(u) = C \tag{36}$$
and
$$\Psi(u) \leq (C + |\widetilde{C}|)e^{-Ru} \tag{37}$$
and thus the Cramér-Lundberg approximation and the Lundberg inequality hold.

We shall now illustrate R, C, $\widetilde{R}$, and $\Psi(u)$ in the case $\alpha = 1$, $\mu = 1$, and $c = 1.2$. We choose $q_2 = 0.2$ which might be reasonable if α_2 is interpreted as the "increased" risk.

REMARK 17. We have, quite naturally, considered the two-state Markov process as a special case of an M-state Markov process. It can, however, also be looked upon as a special case of a Markov process with independent jumps.

Let σ denote the duration between two successive jumps in a Markov process with independent jumps and recall that $E[\sigma] = \int_S \frac{p_L(dy)}{\eta(y)}$. In illustrations we generally put $\alpha = 1$, which means that we use the mean duration between two successive claims as a unit of time. (A strict definition of "the mean duration between two successive claims" will be given in Chapter 5.) The intensity process is generally a model of the fluctuation in the underlying risk, and it "acts" in real time. Therefore $E[\sigma]$ is an important characteristic of the intensity process and shall be related to both the "risk fluctuation" and the volume of the insurance business.

Consider a Markov process with independent jumps and whose jump measure puts its mass at α_1 and α_2. Put

$$p_1 = p_L(\{\alpha_1\}) \quad \text{and} \quad p_2 = p_L(\{\alpha_2\}).$$

It is easy to realize that this is a two-state Markov process with

$$\eta_1 = \eta(\alpha_1)p_2 \quad \text{and} \quad \eta_2 = \eta(\alpha_2)p_1$$

and we have

$$E[\sigma] = \frac{p_1}{\eta_1(\alpha_1)} + \frac{p_2}{\eta_2(\alpha_2)} = \frac{p_1 p_2}{\eta_1} + \frac{p_1 p_2}{\eta_2}.$$

Obviously the representation of a two-state Markov process as a Markov process with independent jumps is not unique. It is natural to choose $p_1 = p_2 = 1/2$ since then $E[\sigma]$ is maximized and thus the number of "false" jumps, i.e., "jumps" from a state into itself, is minimized. Then we have

$$E[\sigma] = \frac{1}{4}\left(\frac{1}{\eta_1} + \frac{1}{\eta_2}\right). \tag{38}$$

Generally an M-state Markov process does not have a representation as a Markov process with independent jumps unless

$$\mathbf{P} = (\mathbf{I} - d(\mathbf{p}))^{-1}(\mathbf{1}\mathbf{p}^T - d(\mathbf{p}))$$

for some distribution **p**.
□

Although the representation of a two-state Markov process as a Markov process with independent jumps is somewhat artificial, it may be natural to use $E[\sigma]$, as given by (38), as a characteristic since we shall later consider other intensity processes with independent jumps.

Now we specify a model by α_1 and $E[\sigma]$. With the choices made we have

$$\alpha_2 = 5 - 4\alpha_1, \quad \eta_1 = \frac{5}{16E[\sigma]}, \quad \eta_2 = 4\eta_1 = \frac{5}{4E[\sigma]}, \quad \Psi(0) = \frac{1}{c} = 0.8333.$$

Consider first the case $E[\sigma] = 10$, which may be a choice with some practical relevance. Then

$$\eta_1 = 0.03125, \quad \eta_2 = 0.125 \quad \text{or} \quad \frac{1}{\eta_1} = 32 \quad \text{and} \quad \frac{1}{\eta_2} = 8.$$

We have computed the renewal case, i.e., $\alpha_1 = 0$, and the Poisson case, i.e., $\alpha_1 = 1$, in the same way as the other cases. In the renewal case we have used $d(1) = 0$, see Remark 15. This case is illustrated in Table 1.

TABLE 1. Two-state Markov intensity and exponentially distributed claims in the case $\alpha = \mu = 1$, $c = 1.2$, $q_2 = 0.2$, and $E[\sigma] = 10$.

α_1	α_2	R	C	$\widetilde{R}$	$\widetilde{C}$	$\Psi(10)$	$\Psi(100)$	$\Psi(1000)$
0.00	5.0	0.0066	0.8333	1.0000	0.0000	0.7803	0.4320	0.0012
0.05	4.8	0.0072	0.8266	0.9594	0.0068	0.7691	0.4020	0.0006
0.10	4.6	0.0079	0.8192	0.9190	0.0141	0.7567	0.3703	0.0003
0.15	4.4	0.0088	0.8113	0.8786	0.0220	0.7430	0.3368	0.0001
0.20	4.2	0.0098	0.8027	0.8384	0.0306	0.7279	0.3017	0.0000
0.25	4.0	0.0110	0.7934	0.7983	0.0399	0.7111	0.2653	0.0000
0.50	3.0	0.0216	0.7335	0.6012	0.0998	0.5912	0.0844	0.0000
0.75	2.0	0.0580	0.6569	0.4171	0.1764	0.3705	0.0020	0.0000
0.80	1.8	0.0746	0.6506	0.3847	0.1827	0.3124	0.0004	0.0000
0.85	1.6	0.0973	0.6611	0.3563	0.1723	0.2546	0.0000	0.0000
0.90	1.4	0.1264	0.7063	0.3366	0.1270	0.2039	0.0000	0.0000
0.95	1.2	0.1550	0.7911	0.3377	0.0422	0.1693	0.0000	0.0000
1.00	1.0	0.1667	0.8333	0.3795	0.0000	0.1574	0.0000	0.0000

The most striking impression of Table 1 is probably the strong influence deviations from the Poisson case have on the ruin probability. A case like $\alpha_1 = 0.75$ and $\alpha_2 = 2$ must be regarded as rather "harmless" but still $\Psi(10)$ is more than twice the corresponding value in the Poisson case.

Certainly the parameter $E[\sigma]$ is important for the behavior of the risk process. In Table 2 we illustrate this for different values of $E[\sigma]$. Since our intention is solely to illustrate the dependence of $E[\sigma]$ we disregard the fact that very small values of $E[\sigma]$ seem unrealistic.

For (very) small values of $E[\sigma]$ the parameters are close to those in the Poisson case. This is not surprising if we think of the representation of the intensity process as a Markov process with independent jumps. The probability of more than one claim between two successive jumps is very small. Therefore the values of $\lambda(t)$ at different claims are "almost" independent and thus N has independent increments. This implies – compare Theorem 2.11 – that N is "almost" a Poisson process.

Large values of $E[\sigma]$ ought to correspond to the case where N is "almost" a mixed Poisson process. If N is a mixed Poisson process we have, compare (20), for $\alpha_2 > c/\mu$:

$$\Psi(u) = q_2 \cdot 1 + q_1 \frac{\alpha_1 \mu}{c} e^{-R_1 u} \tag{39}$$

which formally means, for the chosen parameter values, that

$$R = 0, \quad C = q_2 = 0.2, \quad \widetilde{R} = R_1 = 1 - \frac{\alpha_1}{1.2}, \quad \widetilde{C} = q_1 \frac{\alpha_1 \mu}{c} = \frac{\alpha_1}{6}.$$

4.1 Markovian intensity: Preliminaries 91

TABLE 2. Two-state Markov intensity and exponentially distributed claims in the case $\alpha = \mu = 1$, $c = 1.2$, and $q_2 = 0.2$. $E[\sigma] = \infty$ indicates a mixed Poisson intensity.

α_1	α_2	$E[\sigma]$	R	C	$\tilde{R}$	$\tilde{C}$	$\Psi(10)$
0.00	5	∞	0.0000	0.2	1	0	0.2
0.25	4	∞	0.0000	0.2	0.7917	0.0417	0.2000
0.50	3	∞	0.0000	0.2	0.5833	0.0833	0.2002
0.75	2	∞	0.0000	0.2	0.3750	0.1250	0.2029
1.00	1	∞	0.1667	0.6667	0.1667	0.1667	0.1574
0.00	5	1000	0.0001	0.8333	1.0000	0.0000	0.8328
0.25	4	1000	0.0001	0.7894	0.7917	0.0439	0.7885
0.50	3	1000	0.0002	0.7144	0.5835	0.1189	0.7130
0.75	2	1000	0.0009	0.5569	0.3754	0.2765	0.5585
1.00	1	1000	0.1667	0.8333	0.1729	0.0000	0.1574
0.00	5	100	0.0007	0.8333	1.0000	0.0000	0.8277
0.25	4	100	0.0012	0.7899	0.7923	0.0435	0.7807
0.50	3	100	0.0024	0.7164	0.5852	0.1169	0.6995
0.75	2	100	0.0083	0.5684	0.3793	0.2650	0.5293
1.00	1	100	0.1667	0.8333	0.2144	0.0000	0.1574
0.00	5	1	0.0480	0.8333	1.0000	0.0000	0.5154
0.25	4	1	0.0685	0.8135	0.8432	0.0199	0.4099
0.50	3	1	0.1004	0.8023	0.7128	0.0311	0.2939
0.75	2	1	0.1426	0.8160	0.6402	0.0174	0.1960
1.00	1	1	0.1667	0.8333	0.7068	0.0000	0.1574
0.00	5	0.1	0.1330	0.8333	1.0000	0.0000	0.2205
0.25	4	0.1	0.1457	0.8321	0.9512	0.0013	0.1938
0.50	3	0.1	0.1566	0.8323	0.9229	0.0010	0.1739
0.75	2	0.1	0.1640	0.8331	0.9191	0.0003	0.1616
1.00	1	0.1	0.1667	0.8333	0.9437	0.0000	0.1574
0.00	5	0.01	0.1625	0.8333	1.0000	0.0000	0.1641
0.25	4	0.01	0.1643	0.8333	0.9938	0.0000	0.1612
0.50	3	0.01	0.1656	0.8333	0.9906	0.0000	0.1591
0.75	2	0.01	0.1664	0.8333	0.9906	0.0000	0.1578
1.00	1	0.01	0.1667	0.8333	0.9937	0.0000	0.1574

These values, indicated by $E[\sigma] = \infty$, are given in Table 2.

In the Poisson case there is no obvious decomposition of $C + \tilde{C} = \alpha\mu/c$ in C and $\tilde{C}$ but we have chosen to put $C = q_2\alpha\mu/c$ and $\tilde{C} = q_1\alpha\mu/c$.

Although Table 2 indicates that the values of R and $\tilde{R}$ for large values of $E[\sigma]$ are close to those values in the mixed Poisson case, there is obviously no convergence of the ruin probabilities. This is, however, not to be

expected since, for example,
$$\Psi(0) = \frac{\alpha\mu}{c} \quad \text{and} \quad \Psi(\infty) = 0 \quad \text{in the two-state Markov case}$$
while, see (39), for $\alpha_2 > c/\mu$:
$$\Psi(0) = q_2 + q_1\frac{\alpha_1\mu}{c} \quad \text{and} \quad \Psi(\infty) = q_2 \quad \text{in the mixed Poisson case.}$$
Mathematically this is nothing strange since (compare the survey of weak convergence given in Section 1.2 and the discussion in Section A.4) $X_n \xrightarrow{d} X$ does not imply $\inf_{t\geq 0} X_n(t) \xrightarrow{d} \inf_{t\geq 0} X(t)$.

4.2 The martingale approach

Our basic approach to Cox models, due to Björk and Grandell (1988, pp. 79 - 84), is an extension of Gerber's "martingale approach" used in Section 1.1. Let us therefore recapitulate the main steps in that approach.

Suppose we have a suitable filtration $\mathbf{F}$, a positive $\mathbf{F}$-martingale (or a positive $\mathbf{F}$-supermartingale) M, and an $\mathbf{F}$-stopping time T. Choose $t_0 < \infty$ and consider $t_0 \wedge T$ which is a bounded $\mathbf{F}$-stopping time. Since M is positive, it follows from Theorem 1.14 – compare (1.18) – that
$$M(0) \geq E^{\mathcal{F}_0}[M(t_0 \wedge T)] \geq E^{\mathcal{F}_0}[M(T) \mid T \leq t_0]P^{\mathcal{F}_0}\{T \leq t_0\} \tag{40}$$
and thus
$$P^{\mathcal{F}_0}\{T \leq t_0\} \leq \frac{M(0)}{E^{\mathcal{F}_0}[M(T) \mid T \leq t_0]}. \tag{41}$$
Obviously M must be related to the risk process X. Let therefore X be adapted to $\mathbf{F}$, i.e., $\mathcal{F}_t \supseteq \mathcal{F}_t^X$ for all $t \geq 0$, and $T = T_u$ be the time of ruin, i.e.,
$$T_u = \inf\{t \geq 0 \mid u + X(t) < 0\}.$$
Then $\Psi(u) = E[P^{\mathcal{F}_0}\{T < \infty\}] = P\{T < \infty\}$. Further, we must choose M such that it is possible to find a good lower bound for the denominator in (41).

In the Poisson case we chose
$$M(t) = \frac{e^{-r(u+X(t))}}{E[e^{-rX(t)}]} = \frac{e^{-r(u+X(t))}}{e^{t(\alpha h(r)-rc)}} \tag{42}$$
which, since X has independent increments, was easily shown to be an $\mathbf{F}^X$-martingale. Using $u + X(T_u) \leq 0$ on $\{T_u < \infty\}$, the lower bound was shown to be given by
$$E[M(T_u) \mid T_u \leq t_0]$$
$$\geq E[e^{-T_u(\alpha h(r)-rc)} \mid T_u \leq t_0] \geq \inf_{0\leq t\leq t_0} e^{-t(\alpha h(r)-rc)}. \tag{43}$$

The last step was to let $t_0 \to \infty$. The facts that
$$\inf_{t\geq 0} e^{-t(\alpha h(r)-rc)} = 1 \tag{44}$$
for $r \leq R$, where R is the positive solution of $h(r) = cr/\alpha$, and that $M(0) = e^{-ru}$ led to Lundberg's inequality. *Note that (44) holds for $r \leq R$ and not only for $r < R$.*

Now we consider a Cox model where N is a Cox process with intensity process $\lambda(t)$. The intensity measure Λ is given by
$$\Lambda(t) = \int_0^t \lambda(s)\ ds.$$
A suitable filtration – compare Proposition 2.18 – is **F** given by $\mathcal{F}_t = \mathcal{F}_\infty^\Lambda \vee \mathcal{F}_t^X$. Note that $\mathcal{F}_0 = \mathcal{F}_\infty^\Lambda$.

We shall make strong use of Lemma 2.19 which says that

(i) $N(t)$ has independent increments relative to $\mathcal{F}_\infty^\Lambda$;

(ii) $N(t) - N(s)$ is Poisson distributed with mean $\Lambda(t) - \Lambda(s)$ relative to $\mathcal{F}_\infty^\Lambda$.

It seems very natural to try to find an **F**-martingale "as close as possible" to the one used in the Poisson case. Therefore we consider
$$M(t) = \frac{e^{-r(u+X(t))}}{e^{\Lambda(t)h(r)-trc}}, \tag{45}$$
where we quite simply have replaced αt with $\Lambda(t)$. It follows by almost obvious modifications of (1.17) that M is an **F**-martingale. Due to the importance of this result we give it as a lemma and "repeat" the proof.

LEMMA 18. *The process M, given by (45), is an **F**-martingale where the filtration **F** is given by $\mathcal{F}_t = \mathcal{F}_\infty^\Lambda \vee \mathcal{F}_t^X$.*

PROOF: The fact that N, and thus X, has independent increments relative to $\mathcal{F}_\infty^\Lambda$ is equivalent to that $X(t) - X(s)$, for $s \leq t$, independent of $\mathcal{F}_s$ relative to $\mathcal{F}_0 = \mathcal{F}_\infty^\Lambda$. Since
$$E^{\mathcal{F}_\infty^\Lambda}[e^{-rX(t)}] = e^{-rct} \sum_{k=0}^\infty \frac{\Lambda(t)^k}{k!} e^{-\Lambda(t)} (h(r)+1)^k$$
$$= e^{-rct + \Lambda(t)(h(r)+1) - \Lambda(t)} = e^{\Lambda(t)h(r)-rct}$$
we get, see (1.17),
$$E^{\mathcal{F}_s}[M(t)] = E^{\mathcal{F}_s}\left[\frac{e^{-r(u+X(t))}}{e^{\Lambda(t)h(r)-rct}}\right]$$
$$= E^{\mathcal{F}_s}\left[\frac{e^{-r(u+X(s))}}{e^{\Lambda(s)-rcs}} \cdot \frac{e^{-r(X(t)-X(s))}}{e^{\Lambda(t)-\Lambda(s)-(t-s)rc}}\right]$$
$$= M(s) \cdot E^{\mathcal{F}_s}\left[\frac{e^{-r(X(t)-X(s))}}{e^{\Lambda(t)-\Lambda(s)-(t-s)rc}}\right] = M(s). \quad \blacksquare$$

A lower bound is easily obtained in the same way as in (43):
$$E^{\mathcal{F}_0}[M(T_u) \mid T_u \leq t_0]$$
$$\geq E^{\mathcal{F}_0}[e^{-(\Lambda(T_u)h(r)-rcT_u)} \mid T_u \leq t_0] \geq \inf_{0 \leq t \leq t_0} e^{-\Lambda(t)h(r)+rct}. \qquad (46)$$

Thus we have, see (41),
$$P^{\mathcal{F}_0}\{T_u \leq t_0\} \leq \frac{M(0)}{E^{\mathcal{F}_0}[M(T_u) \mid T_u \leq t_0]} \leq e^{-ru} \sup_{0 \leq t \leq t_0} e^{\Lambda(t)h(r)-rct}$$

and, by taking expectation,
$$P\{T_u \leq t_0\} \leq e^{-ru} E\left[\sup_{0 \leq t \leq t_0} e^{\Lambda(t)h(r)-rct}\right]. \qquad (47)$$

When $t_0 \to \infty$ in (47) we get
$$\Psi(u) \leq C(r)e^{-ru}, \quad \text{where} \quad C(r) = E\left[\sup_{t \geq 0} e^{\Lambda(t)h(r)-rct}\right]. \qquad (48)$$

Like in the Poisson case we now want to choose r as large as possible and we are led to the following definition of the Lundberg exponent.

DEFINITION 19. *The Lundberg exponent R is defined by*
$$R = \sup\{r \mid C(r) < \infty\},$$
where $C(r)$ is given by (48).

Definition 19 may look harmless, but in general it is probably very difficult to actually determine R. Further, we do not know at all if R is the "right" exponent. Disregarding, for the moment, these "practical," i.e., in reality fundamental, questions we have the following generalization of the Lundberg inequality.

THEOREM 20. *For every $\epsilon > 0$ such that $0 < \epsilon < R$ we have*
$$\Psi(u) \leq C(R-\epsilon)e^{-(R-\epsilon)u},$$
where $C(R-\epsilon) < \infty$.

REMARK 21. The condition $\epsilon > 0$ is unpleasant, but it is quite possible, and really natural, that $C(R) = +\infty$. In fact, $C(R) < \infty$ requires a discontinuity in $C(r)$ at $r = R$. In the Poisson case we have such a discontinuity since $C(r) = 1$ for $r \leq R$ and $C(r) = \infty$ for $r > R$.

From (37) we do, however, know that there exists Cox cases where Lundberg's inequality holds with $\epsilon = 0$. In Section 4.5 we shall consider an alternative approach based on another filtration and another martingale, which in certain cases of markovian intensity leads to a Lundberg inequality with $\epsilon = 0$. This does, however, not mean that $C(r)$ has a "mysterious" discontinuity, but that the "lower bound" is obtained differently.
□

We conclude the discussion of the Cox case in this generality with a result that shows that the stationary Cox case is "more dangerous" than the Poisson case. As in Section 3.3, let R_P be the Lundberg exponent in the corresponding Poisson case, i.e., in a Poisson case with the same α, c, and $F(z)$ as in the Cox case.

THEOREM 22. *Suppose that $\lambda(t)$ is stationary with $E[\lambda(t)] \equiv \alpha$. Then $R \leq R_P$.*

PROOF: Choose any $r > R_P$. Then $h(r)\alpha > rc$ and we have
$$C(r) \geq \sup_{t \geq 0} E\left[e^{\Lambda(t)h(r) - rct}\right] \geq \sup_{t \geq 0} \left[e^{t(\alpha h(r) - rc)}\right] = \infty,$$
where the first inequality is trivial and the second follows from Jensen's inequality. Thus $R \leq r$ and the theorem is proved. ∎

4.3 Independent jump intensity

We now consider a class of intensity processes with "independent jumps." Our discussion is based on Björk and Grandell (1988, pp. 84 - 96).

Intuitively an independent jump intensity is a jump process where the jump times form a renewal process and where the value of the intensity between two successive jumps may depend only on the distance between these two jumps. More formally, let Σ_k, $k = 1, 2, \ldots$ denote the epoch of the kth jump of the intensity process and let $\Sigma_0 \stackrel{\text{def}}{=} 0$. Put

$$\begin{aligned} \sigma_n &= \Sigma_n - \Sigma_{n-1} \\ L_n &= \lambda(\Sigma_{n-1}) \end{aligned} \qquad n = 1, 2, 3, \ldots. \tag{49}$$

Here we understand that λ has right-continuous realization so that $\lambda(t) = L_n$ for $\Sigma_{n-1} \leq t < \Sigma_n$. These notations are illustrated in Figure 1.

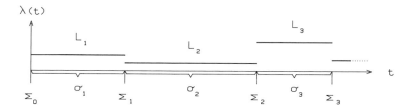

FIGURE 1. Illustration of notation.

96   4 Cox models

DEFINITION 23. *An intensity process* λ *is called*

(i) *an independent jump intensity* if the random vectors
$$(L_1, \sigma_1), (L_2, \sigma_2), (L_3, \sigma_3), \ldots$$
are independent and if $(L_2, \sigma_2), (L_3, \sigma_3), \ldots$ have the same distribution p;

(ii) *an ordinary independent jump intensity* if (L_1, σ_1) also has distribution p;

(iii) *a stationary independent jump intensity* if the distribution of (L_1, σ_1) is chosen such that λ is stationary.

Let (L, σ) be the generic vector for (L_n, σ_n), $n \geq 2$, i.e.,
$$\Pr\{L \in A,\ \sigma \in B\} = p(A \times B) \qquad \text{for } A,\ B \in \mathcal{B}(\mathbf{R}_+).$$
The marginal distribution of L is denoted by p_L, i.e.,
$$p_L(A) = p(A \times \mathbf{R}_+) \qquad \text{for } A \in \mathcal{B}(\mathbf{R}_+).$$

Assume that $E[\sigma] < \infty$ and let q be the distribution of (L_1, σ_1). The following theorem is a consequence of Franken et al. (1981, p. 45).

THEOREM 24. *The intensity* λ *is stationary for*
$$q(A \times B) = \frac{1}{E[\sigma]} \int_B p(A \times (t, \infty))\, dt. \tag{50}$$

Furthermore, for this choice of q,
$$E[f(L_1, \sigma_1)] = \frac{1}{E[\sigma]} E\left[\int_0^\sigma f(L, s)\, ds\right] \tag{51}$$

for any measurable function $f : \mathbf{R}_+^2 \to \mathbf{R}_+$.

INDICATION OF PROOF: The result in Franken et al. (1981, p. 45) is much more general than Theorem 24. Instead of *showing* that Theorem 24 is a consequence we choose to *indicate* the theorem.

It is natural that λ becomes stationary when the renewal process of jumps is stationary. Consider the extended index space $\mathbf{R}$. Let Σ_{-1} be the epoch of the last jump before time 0. Then σ_1, conditioned upon $\Sigma = \Sigma_1 - \Sigma_{-1}$, is uniformly distributed on $[0, \Sigma]$. Using this and (3.1) we get
$$\frac{ds}{E[\sigma]} p(\mathbf{R}_+ \times (s, \infty)) = \frac{ds}{E[\sigma]} \int_s^\infty p(\mathbf{R}_+ \times dy) = P\{\sigma_1 \in (s, s+ds)\}$$
$$= \int_0^\infty P\{\sigma_1 \in (s, s+ds) \mid \Sigma = x\} P\{\Sigma \in (x, x+dx)\}$$
$$= \int_s^\infty \frac{ds}{x} P\{\Sigma \in (x, x+dx)\}$$

and thus
$$P\{\Sigma \in (x, x+dx)\} = \frac{x\,p(\mathbf{R}_+ \times dx)}{E[\sigma]} \tag{52}$$
which is a well-known result from renewal theory.

Further, L_1 is only dependent on Σ, and it depends on Σ in the same way as L depends on σ. Therefore we have
$$P\{L_1 \in (\ell, \ell+d\ell) \mid \Sigma \in (x, x+dx)\} = \frac{p(d\ell \times dx)}{p(\mathbf{R}_+ \times dx)}, \tag{53}$$
where the ratio is interpreted as a Radon-Nikodym derivative. From (52) and (53) we get
$$P\{L_1 \in (\ell, \ell+d\ell),\, \sigma_1 \in (s, s+ds)\}$$
$$= \int_0^\infty P\{L_1 \in (\ell, \ell+d\ell),\, \sigma_1 \in (s, s+ds) \mid \Sigma = x\}\, P\{\Sigma \in (x, x+dx)\}$$
$$= \int_0^\infty P\{L_1 \in (\ell, \ell+d\ell) \mid \Sigma = x\}\, P\{\sigma_1 \in (s, s+ds) \mid \Sigma = x\}\, P\{\Sigma \in (x, x+dx)\}$$
$$= \int_0^\infty \frac{p(d\ell \times dx)}{p(\mathbf{R}_+ \times dx)} \cdot \frac{ds}{x} \cdot \frac{x\,p(\mathbf{R}_+ \times dx)}{E[\sigma]}$$
$$= \frac{ds}{E[\sigma]} \int_s^\infty p(d\ell \times dx) = \frac{ds}{E[\sigma]}\, p(d\ell \times (s, \infty)) \tag{54}$$
which is the differential version of (50).

Further, we have
$$E[f(L_1, \sigma_1)] = \frac{1}{E[\sigma]} \int_0^\infty \int_0^\infty f(\ell, s)\, ds \int_s^\infty p(d\ell \times dx)$$
$$= \frac{1}{E[\sigma]} \int_0^\infty \int_0^\infty \int_0^x f(\ell, s)\, ds\, p(d\ell \times dx)$$
which is the same as (51). ∎

EXAMPLE 25. An interesting special case is when σ, conditioned upon L, is exponentially distributed, since then λ is a Markov process with independent jumps. This means that
$$p(d\ell \times ds) = p_L(d\ell)\, \eta(\ell) e^{-\eta(\ell)s}\, ds \tag{55}$$
and thus
$$E[\sigma] = \int_0^\infty \int_0^\infty p_L(d\ell)\, s\, \eta(\ell) e^{-\eta(\ell)s}\, ds = \int_0^\infty \frac{p_L(d\ell)}{\eta(\ell)}. \tag{56}$$
It follows from (50) that
$$q(d\ell \times ds) = \frac{p(d\ell \times (s, \infty))\, ds}{E[\sigma]} = \frac{p_L(d\ell)}{E[\sigma]} e^{-\eta(\ell)s}\, ds$$

98   4 Cox models

$$= \frac{p_L(d\ell)}{\eta(\ell)E[\sigma]} \eta(\ell)e^{-\eta(\ell)s} \, ds = q_L(d\ell) \, \eta(\ell)e^{-\eta(\ell)s} \, ds, \qquad (57)$$

where q_L is given by (10). Note that we have only required $E[\sigma] < \infty$. We shall return to this example in Section 4.5.
□

In the stationary case it follows from (51) that

$$\alpha = E[L_1] = \frac{E[L\sigma]}{E[\sigma]} \qquad (58)$$

and thus

$$\rho = \frac{cE[\sigma] - \mu E[L\sigma]}{\mu E[L\sigma]}. \qquad (59)$$

In the sequel we assume that $E[\sigma] < \infty$ and $E[L\sigma] < \infty$. We may note that ρ can be interpreted as

$$\frac{\text{the net profit}}{\text{the net cost}} \qquad (60)$$

between two successive jumps is the intensity. Thus (59) is the natural definition of ρ also in the ordinary case.

Certainly Cox processes with independent jump intensity are a very restricted class of Cox models. It is, however, general enough to include a number of non-trivial models while still allowing us to obtain fairly explicit results reasonably simple. Naturally, we shall rely on the renewal structure of the intensity. Like in the renewal case, we shall first consider the ordinary case and then the stationary case and we denote the ruin probabilities by Ψ^0 and Ψ, respectively. First, however, we shall consider an imbedded random walk.

4.3.1 An imbedded random walk

Let N be a Cox process with an ordinary independent jump intensity λ. Then the random variables X_k, $k = 1, 2, \ldots$, defined by

$$X_k \stackrel{\text{def}}{=} -[X(\Sigma_k) - X(\Sigma_{k-1})] \qquad (\Sigma_0 \stackrel{\text{def}}{=} 0) \qquad (61)$$

are independent and identically distributed. Let X be the generic variable for X_k. Then

$$E[X] = -E[X(\Sigma_1)] = -E[E[X(\Sigma_1) \mid L_1, \Sigma_1]]$$
$$= -E[\mu L_1 \Sigma_1 - c\Sigma_1] = -E[\mu L\sigma - c\sigma] = -\rho \mu E[L\sigma] < 0$$

for $\rho > 0$. When nothing else is said we assume that $\rho > 0$.
The process $Y = \{Y_n;\ n = 0, 1, 2, \ldots\}$, where

$$Y_0 = 0 \quad \text{and} \quad Y_n = \sum_{k=1}^n X_k \quad \text{for } n = 1, 2, \ldots \qquad (62)$$

is thus a random walk. $Y_n = -X(\Sigma_n)$ is the loss immediately after the nth jump of the intensity. Put, compare (3.6),

$$\phi(r) \stackrel{\text{def}}{=} E[e^{rX}] = E[e^{-rX(\Sigma_1)}] = E[E[e^{-rc\Sigma_1 + L_1\Sigma_1 h(r)} \mid L_1, \Sigma_1]]$$
$$= E[e^{-rc\sigma + h(r)L\sigma}], \qquad (63)$$

where $h(r)$ is given by Definition 1.3. The function $\phi(r)$ will play the same rôle as $g(r)$ did in the renewal case. There are, however, two important differences.

1. *Ruin occurs between the jumps of the intensity, and therefore a study of this random walk needs, a priori, not be of any relevance.* Its relevance will, however, follow from Section 4.3.2.

2. *It does not follow from Assumption 1.4 that $\phi(r)$ has corresponding regularity properties.*

Since $h(0) = 0$ and $h(r)$ is convex it follows that $\phi(0) = 0$ and – see Lemma 5 in Björk and Grandell (1988, p. 88) – $\phi(r)$ is convex.

EXAMPLE 26. Consider the case where σ is exponentially distributed with mean 1, where L has positive mass arbitrarily far out, i.e.,

$$p_L((\ell, \infty)) > 0 \qquad \text{for all } \ell > 0, \qquad (64)$$

and where σ and L are independent. Then

$$p(d\ell \times ds) = p_L(d\ell) \, e^{-s} \, ds,$$

i.e., – see (55) – the intensity is a markovian jump process with $\eta(\ell) \equiv 1$. Condition (64) holds, for example, when L is exponentially distributed. Thus this example can definitely not be regarded as pathological.

Consider any $r > 0$. Since $h(r) \geq r\mu$ we have, for all ℓ,

$$\phi(r) \geq E[e^{-rc\sigma + r\mu L\sigma}] \geq E[e^{(r\mu\ell - rc)\sigma}]p_\ell((\ell, \infty)).$$

For $r\mu\ell - rc \geq 1$ or for $\ell \geq (1 + rc)/(r\mu)$ (which is the same) we have $E[e^{(r\mu\ell - rc)\sigma}] = \infty$ and thus $\phi(r) = \infty$.
□

The Lundberg exponent R_0 – the subscript 0 will get its explanantion by (68) – is defined by

$$R_0 = \sup\{r \geq 0 \mid \phi(r) \leq 1\} \qquad (65)$$

and the ruin probability $\Psi^{\text{rw}}(u)$ – "rw" stands for "random walk" – by

$$\Psi^{\text{rw}}(u) = P\{\max_{n \geq 1} Y_n > u\}.$$

Obviously $\Psi^{\text{rw}}(u) \leq \Psi^0(u)$.

PROPOSITION 27. *Suppose that $\phi(r) = \infty$ for all $r > 0$. Then, for all $\epsilon > 0$,*

$$\limsup_{u \to \infty} e^{\epsilon u} \Psi^{\text{rw}}(u) = \infty.$$

PROOF: See Proposition 13 in Björk and Grandell (1988, pp. 92 - 93). The idea of that proof is to use the trivial inequality $\Psi^{\text{rw}}(u) \geq P\{Y_1 > u\}$ and then to prove that
$$\limsup_{u \to \infty} e^{\epsilon u} P\{Y_1 > u\} = \infty. \quad \blacksquare$$

It follows from Proposition 27 that we cannot get any Lundberg inequality unless we assume that $\phi(r) < \infty$ for some $r > 0$. Example 26 shows that this is not an "innocent regularity assumption."

PROPOSITION 28. *Assume that $\phi(r) < \infty$ for some $r > 0$ (and that $\rho > 0$). Then $R_0 > 0$.*

PROOF: It follows from Björk and Grandell (1988, p. 90) that
$$\phi'(x) = E\left[\frac{d}{dx} e^{xX}\right] = E\left[Xe^{xX}\right]$$
for $x < r$, and in particular $\phi'(0) = E[X] < 0$. Since $\phi(0) = 0$ the proposition follows. $\blacksquare$

PROPOSITION 29. *Assume that $\phi(r) < \infty$ for some $r > 0$ and that $\phi(R_0) = 1$. Then*
$$\lim_{u \to \infty} e^{R_0 u} \Psi^{\text{rw}}(u) = C^{\text{rw}},$$
where $0 < C^{\text{rw}} < \infty$.

PROOF: This follows from (3.26). $\blacksquare$

In "well-behaved" cases R_0 is determined by $\phi(r) = 1$. The case we want to avoid in (b) below is when $\phi(r)$ has a jump at $r = R_0$.

PROPOSITION 30.

(a)   Suppose $\phi(r) = 1$ for some $r > 0$. Then $r = R_0$.

(b)   Suppose that $R_0 > 0$ and that $\phi(r) < \infty$ for some $r > R_0$. Then $\phi(R_0) = 1$.

(c)   Suppose that $1 < \phi(r) < \infty$ for some $r > 0$. Then $\phi(R_0) = 1$.

PROOF: The result (a) follows from the strict convexity of $\phi(r)$ and (b) is trivial. Choose r such that $1 < \phi(r) < \infty$. From (63) and dominated convergence it follows that
$$\phi(r-) = \lim_{x \uparrow r} \phi(x) = E[\lim_{x \uparrow r} e^{xX}] = \phi(r).$$

Thus $\phi(r-) < 1$ is impossible, and (c) follows by convexity. $\blacksquare$

EXAMPLE 26. CONTINUED. Assume that σ is exponentially distributed with mean 1 and that σ and L are independent but that there exists ℓ_0 such that
$$p_L((\ell, \infty)) = 0 \qquad \text{for } \ell > \ell_0.$$

Then
$$\phi(r) \leq E[e^{-rc\sigma + h(r)\ell_0\sigma}] < \infty \quad \text{for } h(r)\ell_0 - rc < 1$$

and thus it follows from Proposition 28 that $R_0 > 0$. We find it somewhat surprising that the exponential decrease of $\Psi^{rw}(u)$ is "destroyed" as soon as (64) holds, *independently* of how fast $p_L((\ell, \infty)) \searrow 0$ as $\ell \to \infty$.
□

A natural question, under the assumption that $\phi(r) < \infty$ for some $r > 0$, is if R_0 may be defined as "the positive solution of $\phi(r) = 1$." Formally, this is not the case, since "pathological" cases of the kind discussed in connection with Assumption 1.4 can occur. To realize that we can choose $\sigma = s_0$ p-a.s. and let L have a "pathological" distribution. We do not know if there exist any natural examples where $\phi(r)$ has a jump at $r = R_0$. Anyhow, from Proposition 30 it follows that this is no big problem.

4.3.2 Ordinary independent jump intensity

Recall from Section 4.2 that
$$R = \sup\{r \mid C(r) < \infty\}, \quad \text{where} \quad C(r) = E\left[\sup_{t \geq 0} e^{\Lambda(t)h(r) - rct}\right].$$

We shall first consider conditions for $C(r) < \infty$. Since λ is piecewise constant, Λ will be piecewise linear. Thus also $\Lambda(t)h(r) - rct$ will be piecewise linear and it is enough to look at $e^{\Lambda(t)h(r)-rct}$ at the jump times of λ. Formally, we define the discrete time process W and the random variable W^* by
$$W_n = e^{\Lambda(\Sigma_n)h(r) - rc\Sigma_n} \quad \text{and} \quad W^* = \sup_{n \geq 0} W_n. \tag{66}$$

Note that we have suppressed the dependence on r in W and W^*. Thus we have
$$C(r) = E[W^*] \tag{67}$$

and we have reduced the problem of analyzing $E[\sup_{t \geq 0} e^{\Lambda(t)h(r)-rct}]$ to the simpler problem of analyzing $E[\sup_{n \geq 0} W_n]$.

Put $Y_n(r) = L_n \sigma_n h(r) - rc\sigma_n$ and $Y(r) = L\sigma h(r) - rc\sigma$. Then $\phi(r) = E[e^{Y(r)}]$. We shall use the obvious facts that
$$W_n = \exp\left(\sum_{j=1}^n Y_n(r)\right) = \prod_{j=1}^n \exp(Y_n(r))$$

and that $Y_n(r)$, $n = 1, 2 \ldots$, are independent and identically distributed random variables.

PROPOSITION 31. *Suppose that the distribution of $Y(r)$ is not concentrated to one point. Then $\phi(r) < 1$ is a necessary condition for $C(r) < \infty$.*

PROOF: See Proposition 5 in Björk and Grandell (1988, p. 86).

If $\phi(r) > 1$ the result is obvious since $E[W_n] = \phi(r)^n \nearrow \infty$ as $n \to \infty$.

We shall indicate the idea of the proof in the case $\phi(r) = 1$. Since $1 = E[e^{Y(r)}]$ it follows from Jensens' inequality that $E[Y(r)] < 0$ and thus that $W_n \to 0$ P-a.s. as $n \to \infty$.

Assume now that $C(r) < \infty$. Then W is a uniformly integrable martingale (a concept that we have not discussed) and $W_n = E[W_\infty \mid \mathcal{F}^W]$ where $W_n \to W_\infty$ P-a.s. as $n \to \infty$. We have just proved that $W_\infty = 0$ and thus $W_n \equiv 0$, which contradicts $\phi(r) = 1$. ∎

We have not managed to give a sufficent condition for $C(r) < \infty$ in terms of $\phi(r)$. In order to give such a condition we consider $\phi(\delta, r)$, defined by

$$\phi(\delta, r) \stackrel{\text{def}}{=} E[e^{(1+\delta)Y(r)}] = E[e^{(1+\delta)(h(r)L\sigma - rc\sigma)}].$$

By the same arguments as used for $\phi(r)$ ($= \phi(0, r)$) it follows that $\phi(\delta, r)$ is convex.

PROPOSITION 32. *A sufficient condition for $C(r) < \infty$ is that $\phi(\delta, r) < 1$ for some $\delta > 0$.*

PROOF: See Proposition 6 in Björk and Grandell (1988, pp. 86 - 87). The idea of the proof is that $\phi(\delta, r) < 1$ implies that $\{W_n^{(1+\delta)}\}$ is a positive supermartingale. Observing that $\sup_{n \geq 0} W_n^{(1+\delta)} = (W^*)^{(1+\delta)}$ and using a standard martingale inequality it follows that $P\{W^* \geq x\} \leq K x^{-(1+\delta)}$ and that $C(r) < \infty$. ∎

Motivated by Propositions 31 and 32 we define the constants R_δ and R_+ by

$$R_\delta = \sup\{r \geq 0 \mid \phi(\delta, r) \leq 1\} \quad \delta \geq 0, \tag{68}$$

$$R_+ = \sup\{r \geq 0 \mid \phi(\delta, r) \leq 1 \text{ for some } \delta > 0\}. \tag{69}$$

Note that (68) for $\delta = 0$ agrees with (65). The notation R_+ is clarified by the following lemma.

LEMMA 33. *R_δ is non-decreasing as a function of δ, and $\lim_{\delta \downarrow 0} R_\delta = R_+$.*

PROOF: See Lemma 1 in Björk and Grandell (1988, pp. 87 - 88). The idea of the proof is to use that the L_p-norm – on a probability space – is non-decreasing in p. For any random variable ζ the L_p-norm $\|\zeta\|_p = E[|\zeta|^p]^{1/p}$. Using

$$\phi(\delta, r) = E[e^{(1+\delta)Y(r)}] = \|e^{Y(r)}\|_{(1+\delta)}^{(1+\delta)}$$

the lemma follows. ∎

From a computational point of view R_0 is much easer to handle than R. It follows from Propositions 31 and 32 that

$$R_+ \leq R \leq R_0.$$

(Since $\Psi^{\mathrm{rw}}(u) \leq \Psi^0(u)$ the inequality $R \leq R_0$ – almost – also follows from Proposition 29.)

We shall now show that we have in fact $R = R_0$ by proving that $R_+ \geq R_0$. At first this may seem surprising, since we have not managed to give necessary and sufficent conditions for $C(r) < \infty$ in terms of $\phi(r)$. The following theorem is the main result of this section.

THEOREM 34. *Lundberg's inequality (in the version given by Theorem 20) holds with*
$$R = \sup\{r \geq 0 \mid \phi(r) \leq 1\}.$$

PROOF: Although this is Theorem 5 in Björk and Grandell (1988, pp. 88 - 89) we shall give the full proof. If $R_0 = 0$ the result is trivial, so assume that $R_0 > 0$. Choose any $r \in (0, R_0)$. In order to prove that $R_+ \geq R_0$, it is sufficient to show the existence of $\delta > 0$ such that $\phi(\delta, r) \leq 1$. Therefore we choose a $\delta > 0$ small enough to ensure $r' \stackrel{\text{def}}{=} r(1+\delta) < R_0$. Since $\phi(r') \leq 1$, it is enough to show that $\phi(r') \geq \phi(\delta, r)$.

We have
$$\phi(r') - \phi(\delta, r) = E[e^{-r(1+\delta)c\sigma}(e^{-h(r(1+\delta))L\sigma} - e^{-(1+\delta)h(r)L\sigma})].$$
From the convexity of h, together with $h(0) = 0$, we get $h(r(1+\delta)) \geq (1+\delta)h(r)$ and thus
$$\phi(r') - \phi(\delta, r) \geq E[e^{-r(1+\delta)c\sigma}(e^{-(1+\delta)h(r)L\sigma} - e^{-(1+\delta)h(r)L\sigma})] = 0$$
and the theorem follows. ∎

The Lundberg exponent R is the "right" exponent in the following sense.

THEOREM 35. *Assume that $\phi(r) < \infty$ for some $r > R > 0$. Then*
$$\lim_{u \to \infty} e^{(R+\epsilon)u} \Psi^0(u) = \infty$$
for every $\epsilon > 0$.

PROOF: Since $\Psi^{\mathrm{rw}}(u) \leq \Psi^0(u)$ the theorem follows from Propositions 29 and 30. ∎

REMARK 36. If $R = 0$ it follows from Proposition 28 that $\phi(r) = \infty$ for all $r > 0$. Then it follows from Proposition 27 that
$$\limsup_{u \to \infty} e^{\epsilon u} \Psi^0(u) = \infty$$
for every $\epsilon > 0$, so R is (formally) the "right" exponent also in this case. □

4.3.3 Stationary independent jump intensity

Now we consider the case when (L_1, σ_1) has distribution q given by (50). We shall show that Theorem 34 also holds in this case.

THEOREM 37. *Lundberg's inequality (in the version given by Theorem 20) holds with*
$$R = \sup\{r \geq 0 \mid \phi(r) \leq 1\}.$$

PROOF: This is Proposition 1 in Björk and Grandell (1988, pp. 94 - 95), from which the proof is taken.

If $R = 0$ the result is trivial, so assume that $R > 0$. Put
$$\varphi(r) = E[e^{-rc\sigma_1 + h(r)L_1\sigma_1}].$$

It follows from (48) that it is enough to prove that $\varphi(R - \epsilon) < \infty$ for every $\epsilon > 0$.

Choose $\epsilon \in (0, R)$ and put $r = R - \epsilon$. It follows from (51) that
$$E[\sigma]\varphi(r) = E\left[\int_0^\sigma e^{-rcs + h(r)Ls}\, ds\right] \leq E\left[\sigma(1 + e^{-rc\sigma + h(r)L\sigma})\right].$$

Thus it is enough to show that $E[\sigma e^{-rc\sigma + h(r)L\sigma}] < \infty$.

Put $G(r, L) = -rc + h(r)L$. We have $G(r, L) > 0$ if $L > rc/h(r)$. Further, we have
$$\frac{\partial}{\partial r} G(r, L) = -c + h'(r)L.$$

Since $h(0) = 0$ and since h is convex we have $h'(r) > h(r)/r$ and thus
$$\frac{\partial}{\partial r} G(r, L) > -c + \frac{h'(r)rc}{h(r)} = \frac{c(h'(r)r - h(r))}{h(r)} > 0 \quad \text{for } L > rc/h(r).$$

Put $K = c(h'(r)r - h(r))/h(r)$ and choose $r' \in (r, R)$. For $L > rc/h(r)$ we have $G(r', L) - G(r, L) > (r' - r)K$ since $G(r, L)$ is convex in r. Using this we get
$$E[\sigma e^{-rc\sigma + h(r)L\sigma}]$$
$$\leq E\left[\sigma e^{\sigma G(r,L)} I_{\{L \leq rc/h(r)\}}(L)\right] + E\left[\sigma e^{\sigma G(r,L)} I_{\{L > rc/h(r)\}}(L)\right]$$
$$\leq E[\sigma] + E\left[e^{\sigma G(r',L)} \sigma e^{-\sigma(G(r',L) - G(r,L))} I_{\{L > rc/h(r)\}}(L)\right]$$
$$\leq E[\sigma] + \phi(r') \sup_{s \geq 0} s e^{-s(r'-r)K} < \infty,$$

where $I_B(L) = \begin{cases} 1 \text{ if } L \in B \\ 0 \text{ if } L \notin B \end{cases}$. ∎

EXAMPLE 38. Consider the case when $\sigma = s_0$ *p*-a.s. This is essentially the model studied by Ammeter (1948) mentioned in the introduction to this chapter.

If follows from (50) that
$$q(A \times B) = \frac{1}{s_0} \int_0^{s_0} I_B(t)\, dt\, p_L(A)$$

which implies that L_1 and σ_1 are independent, that L_1 has the same distribution as L, and that σ_1 is uniformly distributed on $[0, s_0]$. Certainly this is well-known, and we point it out merely as an illustration of (50).
We have
$$\phi(r) = E[e^{-rcs_0 + h(r)Ls_0}] = e^{-rcs_0} E[e^{h(r)Ls_0}].$$
Following Ammeter we consider the case when L is Γ-distributed with $E[L] = 1$ and $\text{Var}[L] = \beta$. This means that
$$p_L(d\ell) = \frac{\ell^{(1/\beta)-1}}{\beta^{1/\beta} \, \Gamma(1/\beta)} e^{-\ell/\beta} \, d\ell$$
from which we get
$$\phi(r) = e^{-rcs_0} \left(\frac{\beta}{\beta - h(r)s_0} \right)^\beta$$
for r such that $h(r) < \beta/s_0$. Thus R is the positive solution of
$$h(r) = \frac{\beta}{s_0} \left(1 - e^{-rcs_0/\beta} \right),$$
which is eq. (35) in Ammeter (1948, p. 196). Ammeter's derivation is completely different from ours and is based on the infinite divisibility of the Γ-distribution.
□

4.4 Markov renewal intensity

We shall now consider a class of intensity processes which generalizes the independent jump intensity models and includes M-state Markov processes. The discussion is based on Björk and Grandell (1988, pp. 96 - 105).

The formal definition of Markov renewal intensities looks – at least at a first glance – rather horrible, and we believe that the reader wants to feel more convinced about its relevance in order not to leave this section with aversion.

Let us therefore go back to the discussion in Remark 16. There we considered a two-state Markov process and interpreted α_1 as the "normal" risk which now and then was replaced by an "increased" risk α_2. At least two questions appear immediately: why shall the "durations" of the risks be exponentially distributed and why shall the intensities – especially the "increased" intensity – be deterministic? In automobile insurance we interpreted α_2 as the risk under "bad" conditions. Let us, for example, think of "bad" conditions caused by precipitation. It is then reasonable to relate the risk intensity to the rain intensity, which certainly varies from shower to shower. Furthermore, at least in areas with similar climate as Sweden, the durations of dry and wet periods are not exponentially distributed.

DEFINITION 39. An intensity process λ is called a *Markov renewal intensity* if the following holds:

(i) λ is a jump process;

(ii) there is given an irreducible discrete time Markov chain θ with state space $\{1, 2, \ldots, M\}$ and transition matrix $\mathbf{P} = (p_{ij})$;

(iii) when θ makes its kth visit to state i, the intensity λ takes the (stochastic) value L_k^i and stays constant at that value. After a (stochastic) time σ_k^i we let θ change state according to $\mathbf{P}$ and, supposing θ now enters j for the lth time, λ jumps to L_l^j and stays there σ_l^j etc;

(iv) for a given i, the random vectors $\{(L_k^i, \sigma_k^i)\}_{k=1}^{\infty}$ are independent and identically distributed with generic vector (L^i, σ^i);

(v) the random vectors $\{(L_k^i, \sigma_k^i)\}_k^i$ are independent. They are also independent of θ.

The Markov chain $\theta = \{\theta_n;\ n = 1,\ 2, \ldots\}$ may be looked upon as a description of "the risk type." In the above discussion of automobile insurance we may let state 1 correspond to "normal" risk and state 2 to "increased" risk or precipitation period. Then $M = 2$ and $\mathbf{P} = \begin{pmatrix} 0 & 1 \\ 1 & 0 \end{pmatrix}$.

Already in this example it might be natural to consider $M > 2$ if we, for example, separate between rain and snow. In this case it is not enough to consider $M = 3$ – which might be the first idea – since then it is probably completely unrealistic to let θ be a Markov chain. (If we are in a dry period, and know that the precipitation in a preceding period was snow, it is high probability that the precipitation in the next period also will be snow.) If we, however, extend the classification of the "risk types," for example, by including information about the temperature, much realism may be gained.

Conditions (iii) – (v) say that the successive periods are independent, conditioned upon θ, and that their stochastic properties only depend on the "risk type."

For $M = 1$ we are back in independent jump intensities. Strictly speaking we are back in ordinary independent jump intensities, since (iv) implies that (L_k, σ_k) has the same distribution for all k.

If $L^i = \alpha_i$ a.s., σ^i is exponentially distributed with mean $1/\eta_i$, and $p_{ii} = 0$ for all i the intensity process is an M-state Markov process.

Although θ is *defined* as a discrete time process, it can, by the construction above, also be viewed as a continuous time process $\{\theta(t);\ t \geq 0\}$ where $\theta(t)$ is "the risk type" at time t.

REMARK 40. This remark is only about terminology, and may very well be

omitted. We have chosen the name "Markov renewal intensity" since (ii) is a Markov property and (v) is an independence – or renewal – property. The chosen name may be criticized since the intensity is not a Markov renewal process. In fact, a Markov renewal process is a marked point process, see for example Franken et al. (1981, p. 18) or Karr (1986, p. 344), highly related to semi-Markov processes. The process $\theta(t)$ is a semi-Markov process with the special property that the distribution of the time between two successive jumps only depends on the state which the process jumps from. In a general semi-Markov process the time between two successive jumps may depend both on the state which the process jumps from and the state which the process jumps to.
□

Let us now introduce some notation:

Σ_n^i = the time of the nth entrance of $\theta(t)$ to state i;

$W_n^i = e^{\Lambda(\Sigma_n^i) h(r) - rc\Sigma_n^i}$;

$W^{i*} = \sup_{n \geq 0} W_n^i, \quad i = 1, \ldots, M$;

$Y_1^i(r) = \Lambda(\Sigma_1^i) h(r) - rc\Sigma_1^i$;

$Y_k^i(r) = \left(\Lambda(\Sigma_k^i) - \Lambda(\Sigma_{k-1}^i)\right) h(r) - rc\left(\Sigma_k^i - \Sigma_{k-1}^i\right), \quad k = 2, 3, \ldots$.

Observe that the random variables $\{Y_k^i(r)\}_{k=1}^\infty$ are independent and that the random variables $\{Y_k^i(r)\}_{k=2}^\infty$ are, furthermore, identically distributed. Let $Y^i(r)$ denote the generic variable for $Y_k^i(r)$, $k = 2, 3, \ldots$.

From the piecewise linearity of Λ it follows, exactly as in the independent jump case, that $C(r) = E[\max(W^{1*}, \ldots, W^{M*})]$. Since $W^{1*}, \ldots, W^{M*}$ is a finite collection of non-negative random variables we have

$$C(r) < \infty \text{ if and only if } E[W^{i*}] < \infty \text{ for all } i = 1, \ldots, M.$$

Using the fact that $W_n^i = \prod_{k=1}^n \exp(Y_k^i(r))$ we obtain, with exactly the same arguments as in the proof of Proposition 31, the following lemma.

LEMMA 41. *A necessary condition for $C(r) < \infty$ is that*

$$\Phi_{ii} < 1, \quad i = 1, \ldots, M, \tag{70}$$

$$\Phi_{ki} < \infty, \quad k, i = 1, \ldots, M, \tag{71}$$

where

$$\Phi_{ii} = E\left[e^{Y^i(r)}\right], \tag{72}$$

$$\Phi_{ki} = E\left[e^{Y_1^i(r)} \mid \theta(0) = k\right]. \tag{73}$$

(Note that (72) agrees with (73) for $k = i$.) It is also easy to see that we can more or less copy the "δ-reasoning" in Section 4.3.2. Thus we have the following – yet almost useless – analog of Theorem 34.

PROPOSITION 42. *Lundberg's inequality (in the version given by Theorem 20) holds with*

$$R = \sup\{r \geq 0 \mid (70) \text{ and } (71) \text{ are satisfied}\}.$$

The problem is thus to find conditions which ensure that (70) and (71) hold.

Recall the matrix notation introduced in Section 4.1. Let $\mathbf{A} = (a_{ij})$ be an $M \times M$ matrix with eigenvalues $\kappa_1, \ldots, \kappa_M$. (The eigenvalues are solutions of the equation $\det(\mathbf{A} - \kappa \mathbf{I}) = 0$.) The *spectral radius* of $\mathbf{A}$, $\mathrm{spr}(\mathbf{A})$, is defined by

$$\mathrm{spr}(\mathbf{A}) = \max(|\kappa_1|, \ldots, |\kappa_M|).$$

If $\mathbf{A}$ is irreducible and non-negative it follows from Frobenius' theorem that $\mathbf{A}$ has a simple *maximal eigenvalue*, $\kappa[\mathbf{A}]$, such that

$$\kappa[\mathbf{A}] = \mathrm{spr}(\mathbf{A}).$$

To the maximal eigenvalue there corresponds strictly positive eigenvectors. If some component $a_{ij} = \infty$ we define, as a convention, $\kappa[\mathbf{A}] = \infty$.

Let $\boldsymbol{\phi}$ be the vector with components

$$\phi_i(r) = E[e^{-rc\sigma^i + h(r)L^i\sigma^i}] \tag{74}$$

and put $R_\infty = \sup(r \geq 0 \mid \boldsymbol{\phi} < \infty)$.

PROPOSITION 43. *If* $\mathrm{spr}(d(\boldsymbol{\phi})\mathbf{P}) < 1$ *then (70) and (71) hold.*

PROOF: This is Proposition 6 in Björk and Grandell (1988, pp. 98 - 99), from which the proof is taken.

We have

$$\Phi_{ki} = \phi_k(r)p_{ki} + \sum_{j \neq i} \phi_k(r)p_{kj}\Phi_{ji} \tag{75}$$

or

$$\Phi_{ki} = q_{ki} + \sum_{j \neq i} q_{kj}\Phi_{ji} = \sum_{j=1}^{M} q_{kj}\Phi_{ji} + q_{ki}(1 - \Phi_{ii}), \tag{76}$$

where $\mathbf{Q} = (q_{ki}) = d(\boldsymbol{\phi})\mathbf{P} = (\phi_k(r)p_{ki})$. In particular (76) implies

$$\Phi_{ii} = \sum_{j=1}^{M} q_{ij}\Phi_{ji} + q_{ii}(1 - \Phi_{ii}). \tag{77}$$

Put $\boldsymbol{\Phi} = (\Phi_{ki})$. Then it follows from (76) that

$$\boldsymbol{\Phi} = \mathbf{Q}\boldsymbol{\Phi} + \mathbf{Q}d(\mathbf{I} - \boldsymbol{\Phi}) \tag{78}$$

or

$$(\mathbf{I} - \mathbf{Q})\boldsymbol{\Phi} = \mathbf{Q}d(\mathbf{I} - \boldsymbol{\Phi}). \tag{79}$$

$\mathrm{spr}(\mathbf{Q}) < 1$ implies that $\det(\mathbf{I} - \mathbf{Q}) \neq 0$ and that

$$\sum_{n=0}^{\infty} \mathbf{Q}^n = (\mathbf{I} - \mathbf{Q})^{-1}. \tag{80}$$

It follows from (79) that $\Phi = (I - Q)^{-1}Qd(I - Q)$ and thus

$$d(\Phi) = d((I - Q)^{-1}Q)d(I - Q). \tag{81}$$

Equation (80) implies that $(I - Q)^{-1}$ is non-negative and thus it follows from (81) that $\Phi_{ii} = b_i(1 - \Phi_{ii})$ for some $b_i \geq 0$. Thus $\Phi_{ii} = b_i/(1 + b_i) < 1$, i.e., (70) holds.

We will now show that (70) implies (71). Let i and j be fixed and assume that $\Phi_{ji} = \infty$. It follows from (75) that $\Phi_{ki} = \infty$ for all k thus that $p_{kj} > 0$. Since $\mathbf{P}$ is irreducible $p_{kj} > 0$ for some $k \neq j$. For such a k it follows from (77) that $\Phi_{kk} = \infty$. ∎

From Propositions 42 and 43 we get the following theorem, which is the main result of this section.

THEOREM 44. *Lundberg's inequality (in the version given by Theorem 20) holds with*

$$R = \sup\{r \geq 0 \mid \operatorname{spr}(d(\phi)\mathbf{P}) < 1\}.$$

We shall now show that R – given by Theorem 44 – is the "right" exponent in the sense of Theorem 35. The crucial step in the proof is following lemma, which can be looked upon as a converse of Proposition 43.

LEMMA 45. *If $r \neq R_\infty$ then $\Phi_{ii} < 1$ for some i implies $\operatorname{spr}(d(\phi)\mathbf{P}) < 1$.*

PROOF: This is Lemma 8 in Björk and Grandell (1988, pp. 100 - 101). The full proof is rather technical, and we content ourselves with sketching it in the case when $\mathbf{Q}$ is aperiodic and has distinct eigenvalues. Without loss of generality we assume that $\Phi_{11} < 1$.

For $r > R_\infty$ we have $\phi_k(r) = \infty$ for some k. Using the irreducibility of $\mathbf{P}$ if follows that $\Phi_{11} = \infty$ which is in agreement with the convention $\operatorname{spr}(\mathbf{Q}) = \infty$.

For $r < R_\infty$ it follows from (78) that

$$\Phi = Q\Phi + Qd(I - \Phi) = Q^2\Phi + Q^2d(I - \Phi) + Qd(I - \Phi) = \ldots$$

$$\ldots = Q^N\Phi + \sum_{k=1}^{N} Q^k d(I - \Phi) \geq \sum_{k=1}^{N} Q^k d(I - \Phi).$$

Let $\kappa_1, \ldots, \kappa_M$ denote the (distinct) eigenvalues of $\mathbf{Q}$ and put $\kappa_1 = \operatorname{spr}(Q)$. Then (since $\mathbf{Q}$ is aperiodic) $|\kappa_j| < \operatorname{spr}(Q)$ for $j = 2, \ldots, M$. By spectral decomposition of $\mathbf{Q}$ it follows that

$$\Phi \geq \sum_{k=1}^{N}\left(\sum_{j=1}^{M} \kappa_j^k S_j\right) d(I - \Phi),$$

where $\mathbf{S}_1$ is a strictly positive matrix. Thus in particular

$$\Phi_{11} \geq \sum_{k=1}^{N}(Q^k)_{11}(1 - \Phi_{11}) = \sum_{k=1}^{N}\left(\sum_{j=1}^{M} \kappa_j^k (S_j)_{11}\right)(1 - \Phi_{11}).$$

Thus we must have

$$\sum_{k=1}^{N} (\mathbf{Q}^k)_{11} \leq \Phi_{11}/(1 - \Phi_{11}) < \infty$$

and a sufficient condition for this is that

$$(\mathbf{Q}^k)_{11} = \mathrm{spr}(\mathbf{Q})^k (\mathbf{S}_1)_{11} + \sum_{j=2}^{M} \kappa_j^k (\mathbf{S}_j)_{11} \to 0 \quad \text{as} \quad k \to \infty$$

which holds if and only if $\mathrm{spr}(Q) = \mathrm{spr}(d(\phi)\mathbf{P}) < 1$. ∎

Let $\Psi_i(u)$ denote the ruin probability when $\theta(0) = i$, i.e., when a period of "risk type" i starts at time 0. We have the following correspondence of Theorem 35.

THEOREM 46. *Assume that $\Phi_{ii}(r) < \infty$ for some $r > R > 0$. Then*

$$\lim_{u \to \infty} e^{(R+\epsilon)u} \Psi_i(u) = \infty$$

for every $\epsilon > 0$.

PROOF: Put $X_k = -[X(\Sigma_k^i) - X(\Sigma_{k-1}^i)]$. Observe that $X_1, X_2, \ldots$ are independent and identically distributed and that, compare (63), $\Phi_{ii}(r) = E[e^{rX_k}]$. From Lemma 45 and Proposition 30 (b) it follows that $\Phi_{ii}(R) = 1$ and the theorem follows from Proposition 29 with ϕ replaced by Φ_{ii}, in the same way as Theorem 35 followed. ∎

As in the case of independent jump intensity, we can consider the stationary case.

Let p^i denote the distribution of (L^i, σ^i), i.e.,

$$p^i(A \times B) = P\{L^i \in A, \sigma^i \in B\},$$

and let, compare (50), (L_s^i, σ_s^i) – s stands for "stationary" – be a random vector with distribution

$$q^i(A \times B) = \frac{1}{E[\sigma^i]} \int_B p^i(A \times (t, \infty)) \, dt.$$

Let $\mathbf{p} = (p_k)$ be the stationary distribution of the discrete time Markov chain θ, i.e., the unique solution of $\mathbf{pP} = \mathbf{p}$.

It follows from Franken et al. (1981, p. 33) that the intensity becomes stationary if $\theta(0)$ has distribution $\mathbf{q} = (q_i)$ where

$$q_i = \frac{p_i E[\sigma^i]}{\sum_{j=1}^{M} p_j E[\sigma^j]}$$

and if, conditioned upon $\theta(0) = i$,

$$\lambda(t) = L_s^i \quad \text{for} \quad 0 \leq t < \sigma_s^i.$$

(After σ_s^i the intensity behaves according to Definition 39.) Then we have, compare (58) and (59),

$$E[\lambda(t)] = E[\lambda(0)] = \sum_{i=1}^{M} q_i E[L_s^i] = \frac{\sum_{i=1}^{M} p_i E[\sigma^i] E[L_s^i]}{\sum_{i=1}^{M} p_i E[\sigma^i]} = \frac{\sum_{i=1}^{M} p_i E[L^i \sigma^i]}{\sum_{i=1}^{M} p_i E[\sigma^i]} \qquad (82)$$

and

$$\rho = \frac{\sum_{i=1}^{M} p_i (c E[\sigma^i] - \mu E[L^i \sigma^i])}{\sum_{i=1}^{M} p_i \mu E[\sigma^i]}. \qquad (83)$$

With very similar arguments (as in the proof of Theorem 37) it follows that Theorem 44 also holds in the stationary case.

EXAMPLE 47. A simple and natural case is when $M = 2$ and $\mathbf{P} = \begin{pmatrix} 0 & 1 \\ 1 & 0 \end{pmatrix}$, which is the "Markov renewal generalization" of the two-state Markov intensity. It is then natural – compare Remark 16 – to interpret L^1 as the "normal" risk which now and then is replaced by the "increased" risk L^2. It might by natural to assume $L^1 = \alpha_1$ a.s., but the possibility of a random variation of the intensity between different periods of "increased" risk seems highly relevant.

Then we have

$$\det(d(\phi)\mathbf{P} - \kappa \mathbf{I}) = \det \begin{pmatrix} -\kappa & \phi_1 \\ \phi_2 & -\kappa \end{pmatrix} = \kappa^2 - \phi_1 \phi_2$$

and thus $d(\phi)\mathbf{P}$ has eigenvalues $\pm\sqrt{\phi_1 \phi_2}$ and especially

$$\text{spr}(d(\phi)\mathbf{P}) = \sqrt{\phi_1 \phi_2}.$$

Thus – compare Theorem 44 – the definition of R is simplified to

$$R = \sup\{r \geq 0 \mid \phi_1 \phi_2 < 1\}. \qquad (84)$$

□

REMARK 48. It may be natural – compare Remark 16 – to let the claim distribution F depend on the "risk type," i.e., to let the risk distribution be F_i during a period of "risk type" i. With the obvious modification

$$\phi_i(r) = E[e^{-rc\sigma^i + h_i(r) L^i \sigma^i}],$$

where (of course)

$$h_i(r) = \int_0^\infty e^{rz} \, dF_i(z) - 1,$$

of (74), the derivations go through.
□

4.5 Markovian intensity

Now we continue the discussion of markovian intensities. The discussion in Sections 4.5.1 and 4.5.2 is based on Björk and Grandell (1988, pp. 102 - 110).

4.5.1 Application of the basic approach

In this section we shall – in a routine way – apply the results in Sections 4.2 – 4.4 to

(a)   independent jump markovian intensity;

(b) M-state markovian intensity.

Recall that Lundberg's inequality (in the version given by Theorem 20) says, for $0 < \epsilon < R$, that

$$\Psi(u) \leq C(R-\epsilon)e^{-(R-\epsilon)u},$$

where $C(R-\epsilon) < \infty$. We have, for an independent jump intensity,

$$R = \sup\{r \geq 0 \mid \phi(r) \leq 1\}, \quad \text{where } \phi(r) = E[e^{-rc\sigma + h(r)L\sigma}]$$

and, for a Markov renewal intensity,

$$R = \sup\{r \geq 0 \mid \mathrm{spr}(d(\phi)\mathbf{P}) < 1\}.$$

An independent jump intensity is markovian if (55) holds, i.e., if

$$p(d\ell \times ds) = p_L(d\ell)\, \eta(\ell)e^{-\eta(\ell)s}\, ds.$$

PROPOSITION 49. $\phi(r) \leq 1$ if and only if

(a) $rc + \eta(\ell) - h(r)\ell > 0, \quad p_L\text{-a.s.},$

(b) $\displaystyle\int_0^\infty \frac{\eta(\ell)}{rc + \eta(\ell) - h(r)\ell}\, p_L(d\ell) \leq 1.$

PROOF: It follows from (55) that

$$\phi(r) = \int_0^\infty \int_0^\infty e^{-s[rc+\eta(\ell)-h(r)\ell]}\, \eta(\ell)\, ds\, p_L(d\ell). \tag{85}$$

Since

$$\int_0^\infty e^{-s[rc+\eta(\ell)-h(r)\ell]}\eta(\ell)\, ds$$

$$= \begin{cases} \frac{\eta(\ell)}{rc+\eta(\ell)-h(r)\ell} & \text{for } rc + \eta(\ell) - h(r)\ell > 0 \\ +\infty & \text{for } rc + \eta(\ell) - h(r)\ell \leq 0 \end{cases} \tag{86}$$

it follows that the conditions are sufficient.

If (a) holds it is obvious that (b) is necessary. If (a) does not hold it follows that the integral in (86) is infinite on a set of positive p_L-probability and thus $\phi(r) = \infty$ and thus the conditions are necessary. ∎

A Markov renewal intensity is an M-state markovian intensity if $L^i = \alpha_i$ a.s., σ^i is exponentially distributed with mean $1/\eta_i$ and $p_{ii} = 0$ for all i. It is natural to characterize an M-state Markov process by its intensity matrix $\mathbf{H}$. Recall from (11) that

$$\mathbf{H} = d(\boldsymbol{\eta})(\mathbf{P} - \mathbf{I}) \quad \text{or} \quad \mathbf{P} = d(\boldsymbol{\eta})^{-1}(\mathbf{H} + d(\boldsymbol{\eta})).$$

It follows from Proposition (86) that

$$\phi_i(r) = \frac{\eta_i}{rc + \eta_i - h(r)\alpha_i} \quad \text{for } r < R_\infty, \tag{87}$$

where $R_\infty = \sup\{r \geq 0 \mid rc + \eta_i - h(r)\alpha_i > 0,\ i = 1, \ldots, M\}$. Thus, for $r < R_\infty$,

$$d(\boldsymbol{\phi})\mathbf{P} = [rc\mathbf{I} + d(\boldsymbol{\eta}) - h(r)d(\boldsymbol{\alpha})]^{-1}[\mathbf{H} + d(\boldsymbol{\eta})].$$

Thus Lundberg's inequality (in the version given above) holds with

$$R = \sup\{r \geq 0 \mid \mathrm{spr}([rc\mathbf{I} + d(\boldsymbol{\eta}) - h(r)d(\boldsymbol{\alpha})]^{-1}[\mathbf{H} + d(\boldsymbol{\eta})]) < 1\}. \tag{88}$$

EXAMPLE 50. We now consider the case $M = 2$, i.e., the case where the intensity is a two-state Markov process. Then, see (83),

$$\rho = \frac{\sum_{i=1}^{2} \frac{1}{2}(c\frac{1}{\eta_i} - \mu\alpha_i\frac{1}{\eta_i})}{\sum_{i=1}^{2} \frac{1}{2}\mu\frac{1}{\eta_i}} = \frac{c(\eta_1 + \eta_2) - \mu(\alpha_1\eta_2 + \alpha_2\eta_1)}{\mu(\eta_1 + \eta_2)}$$

and thus positive safety loading is the same as $c > (\alpha_1\eta_2 + \alpha_2\eta_1)/(\eta_1 + \eta_2)$.

It follows from (88) – or easier from (84) and (87) – that

$$R = \inf\{r > 0 \mid B(r) = 1\},$$

where

$$B(r) = \left(\frac{\eta_1}{rc + \eta_1 - h(r)\alpha_1}\right)\left(\frac{\eta_2}{rc + \eta_2 - h(r)\alpha_2}\right). \tag{89}$$

For $h(r) = \mu r/(1 - \mu r)$, i.e., when the claims are exponentially distributed, routine calculations show that the equation $B(r) = 1$ is equivalent to (30) when $\alpha_1 > 0$ and to (26) when $\alpha_1 = 0$.
□

4.5.2 An alternative approach

We shall now consider an alternative to the martingale approach which will allow us to settle some questions when we can use the Lundberg inequality with $\epsilon = 0$.

Recall Proposition 31, which says – in the independent jump case – that $\phi(R) = 1$ implies $C(R) = \infty$ as soon as $Y(r) = L\sigma h(r) - rc\sigma$ is non-deterministic. Thus – compare the discussion in Remark 21 – the Poisson case is probably almost the only case where Theorem 20 holds with $\epsilon = 0$.

As pointed out in Remark 21 we know from (37) that there exist Cox cases where Lundberg's inequality holds with $\epsilon = 0$. In the derivation of (37) we used a "backward differential argument" which essentially means that we took the different possible changes in *both* the intensity *and* the risk process into account. Basic in that derivation was that the vector process $(X, \lambda) = \{(X(t), \lambda(t));\ t \geq 0\}$ – and not only λ – is markovian.

In the martingale approach we used the filtration **F** given by
$$\mathcal{F}_t = \mathcal{F}_\infty^\Lambda \vee \mathcal{F}_t^X$$
which means that the variation of $\lambda(t)$ was considered as already completely known at time $t = 0$.

A way to "combine" these approaches may be to consider the filtration **F** given by
$$\mathcal{F}_t \stackrel{\text{def}}{=} \mathcal{F}_t^\lambda \vee \mathcal{F}_t^X = \mathcal{F}_t^{(X,\ \lambda)} \tag{90}$$
and to base the analysis on some suitable **F**-martingale. The way to find such an **F**-martingale is to apply Proposition 5.

Let H be the generator of the intensity process – see Lemma 7 – and G_α the generator of the classical risk process (with intensity α) – see Lemma 8.

Consider now the vector process (X, λ), with state space $\mathbb{S} \subseteq \mathbb{R}^2$, and denote its generator by A. Thus A acts on functions $v = v(x, \ell)$.

LEMMA 51. *The generator, A, of (X, λ) is given by*
$$(Av)(x, \ell) = (G_\ell v)(x, \ell) + (Hv)(x, \ell),$$
where G_ℓ operates on the x-variable and H on the ℓ-variable.

PROOF: This is Proposition 6 in Björk and Grandell (1988, p. 107).

Since $X(0) = 0$ by definition, we consider, like in the proof of Lemma 8, $Y(t) = y + X(t)$, which has the same generator. We have
$$E[v(Y(\Delta), \lambda(\Delta)) \mid Y(0) = y,\ \lambda(0) = \ell] - v(y, \ell)$$
$$= E[v(Y(\Delta), \lambda(\Delta)) - v(Y(\Delta), \ell) \mid Y(0) = y,\ \lambda(0) = \ell]$$
$$= E[v(Y(\Delta), \ell) - v(y, \ell) \mid Y(0) = y,\ \lambda(0) = \ell].$$

We have, since Y has right-continuous trajectories,
$$\frac{1}{\Delta} E[v(Y(\Delta), \lambda(\Delta)) - v(Y(\Delta), \ell) \mid Y(0) = y,\ \lambda(0) = \ell]$$
$$= E[(Hv)(Y(\Delta), \ell) \mid Y(0) = y,\ \lambda(0) = \ell] + o(1) \to (Hv)(y, \ell)$$

and, since Y and λ have no common jumps,

$$\frac{1}{\Delta} E[v(Y(\Delta), \ell) - v(y, \ell) \mid Y(0) = y, \ \lambda(0) = \ell] \to (G_\ell v)(y, \ell)$$

and the lemma follows. ∎

It now follows from Proposition 5 that $M(t) = v(X(t), \lambda(t))$ for v such that
$$Av \equiv 0 \qquad (91)$$
is an F-martingale. We will, however, not study the martingale equation (91) in full generality. Instead, since we want to apply (40) and since we want to obtain exponential estimates, we restrict ourselves to functions v of the form
$$v(x, \ell) = g(\ell) e^{-r(u+x)},$$
where g is a positive function.

Like in Section 4.1 we let $\Psi_\ell(u)$ denote the ruin probability when $\lambda(0) = \ell$, i.e., when $p_0 = \delta_\ell$. Similarly E_ℓ denotes the expectation operator in that case. Thus we have, for any initial distribution, $E_\ell[\ \cdot\ ] = E[\ \cdot\ \mid \lambda(0) = \ell]$.

We can now formulate the version of Lundberg's inequality, where $\epsilon = 0$. Let, like before, T_u be the time of ruin.

PROPOSITION 52. *Let λ be a markovian jump process, and suppose that $g : \mathbf{R}_+ \to \mathbf{R}_+$ and $R > 0$ satisfy*
$$g(\ell)[\ell h(R) - Rc] + \eta(\ell) \int_0^\infty g(z) p_L(\ell, dz) - \eta(\ell) g(\ell) \equiv 0. \qquad (92)$$

Then
$$\Psi_\ell(u) \le \frac{g(\ell) e^{-Ru}}{E_\ell[g(\lambda(T_u)) \mid T_u < \infty]}. \qquad (93)$$

PROOF: We get, from Lemma 8 since G_ℓ operates on the x-variable,
$$(G_\ell v)(x, \ell) = g(\ell) e^{-r(u+x)} \left(-rc + \ell \int_0^\infty e^{rz} \, dF(z) - \ell \right)$$
$$= g(\ell) e^{-r(u+x)} (\ell h(r) - rc)$$
and, from Lemma 7 since H operates on the ℓ-variable,
$$(Hv)(x, \ell) = e^{-r(u+x)} \left(\eta(\ell) \int_0^\infty g(z) p_L(\ell, dz) - \eta(\ell) g(\ell) \right)$$
and thus (92) is equivalent – Lemma 51 – to (91). This mean that
$$M(t) = g(\lambda(t)) e^{-r(u+X(t))}$$
is an F-martingale and (93) follows from (40). ∎

Now we consider the case when λ is an independent jump markovian intensity. This means – Definition 11 – that $p_L(y, B) = p_L(B)$. As before we denote the ruin probability with Ψ^0 in the ordinary case and with Ψ in the stationary case. Thus we have

$$\Psi^0(u) = \int_0^\infty \Psi_\ell(u)\, p_L(d\ell)$$

and

$$\Psi(u) = \int_0^\infty \Psi_\ell(u)\, q_L(d\ell), \quad \text{where} \quad q_L(d\ell) = \frac{p_L(d\ell)}{\eta(\ell)E[\sigma]}.$$

THEOREM 53. *Let λ be an independent jump markovian process, and suppose there exist $R > 0$ and $\beta > 0$ such that*

(a) $\qquad Rc + \eta(\ell) - h(R)\ell > 0, \qquad p_L\text{-a.s.};$

(b) $\qquad \displaystyle\int_0^\infty \frac{\eta(\ell)}{Rc + \eta(\ell) - h(R)\ell}\, p_L(d\ell) = 1;$

(c) $\qquad \eta(\ell) \geq \beta, \qquad p_L\text{-a.s.}$

Then we have the Lundberg inequalities

$$\Psi_\ell(u) \leq \frac{\eta(\ell)}{Rc + \eta(\ell) - h(R)\ell}\left(1 + \frac{Rc}{\beta}\right)e^{-Ru},$$

$$\Psi^0(u) \leq \left(1 + \frac{Rc}{\beta}\right)e^{-Ru},$$

and

$$\Psi(u) \leq \frac{1}{\beta E[\sigma]}\left(1 + \frac{Rc}{\beta}\right)e^{-Ru}.$$

PROOF: This is Theorem 8 in Björk and Grandell (1988, pp. 108 - 109).

Define g by

$$g(\ell) = \frac{\eta(\ell)}{Rc + \eta(\ell) - h(R)\ell}, \tag{94}$$

which due to (a) is positive. From (b) it follows that g satisfies (92). Thus (93) holds. It follows from (c) that

$$\frac{1}{g(\ell)} = \frac{Rc + \eta(\ell) - h(R)\ell}{\eta(\ell)} \leq \frac{Rc + \eta(\ell)}{\eta(\ell)} \leq 1 + \frac{Rc}{\beta}$$

and thus

$$\frac{1}{E_\ell[g(\lambda(T_u)) \mid T_u < \infty]} \leq 1 + \frac{Rc}{\beta}$$

and the inequality for Ψ_ℓ follows. The inequality for Ψ^0 follows from (b). Similarly we get

$$\Psi(u) \leq \int_0^\infty \frac{\eta(\ell)}{Rc + \eta(\ell) - h(R)\ell} \frac{p_L(d\ell)}{\eta(\ell)E[\sigma]}$$

and the inequality for Ψ follows from (b) and (c). ∎

Except for the "additional" condition (c) Theorem 53 is very close to Proposition 49. Condition (c) "prevents" the intensity from remaining "too long" in a state.

Now we consider the case when λ is an M-state markovian intensity with intensity matrix $\mathbf{H}$. Here the function g in Proposition 52 is simply a vector $\mathbf{g} = (g_i) > \mathbf{0}$, i.e., $g_i > 0$ for $i = 1, \ldots, M$.

THEOREM 54. *Let λ be an M-state markovian intensity, and suppose there exist an $R > 0$ and a vector $\mathbf{g} > \mathbf{0}$ such that*

$$[h(R)d(\boldsymbol{\alpha}) - Rc\mathbf{I} + \mathbf{H}]\mathbf{g} = \mathbf{0}. \tag{95}$$

Then, for any initial distribution, we have

$$\Psi(u) \leq Ce^{-Ru}.$$

PROOF: Using

$$\mathbf{H} = d(\boldsymbol{\eta})(\mathbf{P} - \mathbf{I})$$

it is easily realized that (95) is the vector - matrix version of (92). Since C can be chosen as

$$\frac{\max(g_1, \ldots, g_M)}{\min(g_1, \ldots, g_M)}$$

the theorem follows. ∎

In order to apply Theorem 54 we thus have to proceed in two steps.

(i) Find a positive solution, R, to the equation

$$\det[h(R)d(\boldsymbol{\alpha}) - Rc\mathbf{I} + \mathbf{H}] = 0; \tag{96}$$

(ii) check if

$$\ker[h(R)d(\boldsymbol{\alpha}) - Rc\mathbf{I} + \mathbf{H}] \cap (0, \infty)^M \neq \emptyset.$$

REMARK 55. Formally (95) and (88) look rather different. If, however, there exists an R such that – see (88) –

$$\mathrm{spr}([Rc\mathbf{I} + d(\boldsymbol{\eta}) - h(R)d(\boldsymbol{\alpha})]^{-1}[\mathbf{H} + d(\boldsymbol{\eta})]) = 1,$$

then it follows from Frobenius' theorem that there exists an eigenvector $\mathbf{g} \in \mathbf{R}_+^M$ such that

$$[Rc\mathbf{I} + d(\boldsymbol{\eta}) - h(R)d(\boldsymbol{\alpha})]^{-1}[\mathbf{H} + d(\boldsymbol{\eta})]\mathbf{g} = \mathbf{g}$$

or

$$[\mathbf{H} + d(\boldsymbol{\eta})]\mathbf{g} = [Rc\mathbf{I} + d(\boldsymbol{\eta}) - h(R)d(\boldsymbol{\alpha})]\mathbf{g}$$

from which (95) immediately follows.
□

REMARK 56. Define, for notational reasons, the matrix $\mathbf{H}(r)$ by

$$\mathbf{H}(r) = h(r)d(\boldsymbol{\alpha}) - rc\mathbf{I} + \mathbf{H}. \tag{97}$$

Since – this follows from Asmussen (1989, p. 79) – the matrix $e^{\mathbf{H}(r)}$ is non-negative it follows from Frobenius' theorem that step (ii) above is equivalent

EXAMPLE 50. CONTINUED. For $M = 2$ we have $\mathbf{H} = \begin{pmatrix} -\eta_1 & \eta_1 \\ \eta_2 & -\eta_2 \end{pmatrix}$ and (95) is reduced to

$$\left.\begin{array}{r}(h(R)\alpha_1 - Rc - \eta_1)g_1 + \eta_1 g_2 = 0 \\ \eta_2 g_1 + (h(R)\alpha_2 - Rc - \eta_2)g_2 = 0\end{array}\right\}. \quad (98)$$

Since (96) implies, compare (89),

$$(h(R)\alpha_1 - Rc - \eta_1)(h(R)\alpha_2 - Rc - \eta_2) = \eta_1 \eta_2$$

a solution of (98) is given by, compare (94) and Remark 17,

$$g_k = \frac{2\eta_k}{Rc + 2\eta_k - h(R)\alpha_k}. \quad (99)$$

Comparing the arguments leading to (1.23) and (40) it is not difficult to realize that

$$\Psi_k(u) = \frac{g_k e^{-Ru}}{E_k[g_{\theta(T_u)} e^{-R(u+X(T_u))} \mid T_u < \infty]},$$

where the notation is – hopefully – obvious. Assume now that the claims are exponentially distributed with mean μ. A similar argument as in the continuation of Example 1.7 leads to

$$\Psi_k(u) = \frac{g_k(1 - \mu R)e^{-Ru}}{g_1 P_k\{\lambda(T_u) = \alpha_1 \mid T_u < \infty\} + g_2 P_k\{\lambda(T_u) = \alpha_2 \mid T_u < \infty\}}. \quad (100)$$

If $\alpha_1 = 0$ we are in the renewal case and $\Psi_2(u)$ is reduced to (3.10) since ruin can only occur at a claim which means that

$$P_2\{\lambda(T_u) = \alpha_2 \mid T_u < \infty\} = 1.$$

(Note – by similar arguments – that $\Psi_2(u) = \Psi^0(u)$ in this case.) In the case $\alpha_1 > 0$ (36) ought to follow from (100). We must admit that we have no idea how – or if – this can be proved.
□

The main advantage of the "alternative approach" is, of course, that we can prove Lundberg inequalities without the unpleasant condition "$\epsilon > 0$." The main drawbacks are that we have to assume that λ is markovian and that the detailed analysis in Section 4.3 seems impossible to carry through. In particular it may well happen that no $R > 0$ satisfies (a) and (b) in Theorem 53, compare the continuation of Example 26, in which case the "alternative approach" leaves us completely stranded. The "extension" of this approach, discussed in Section A.3, may in this respect be of some consolation.

As mentioned in the introduction to this chapter, Asmussen (1989) considers the Cramér-Lundberg approximation when the intensity process is an M-state Markov process. His methods are based on Wiener-Hopf factorization, and thus entirely different from ours. As a byproduct he obtains an estimate for the constant C in Theorem 54. Albeit we do not enter into his results, we strongly recommend the reader to consult his paper.

4.6 Numerical illustrations

We will now (by five simple examples) numerically illustrate how the R-values in the Cox case differ from the R_P-values in the Poisson case by computing the ratio R_P/R. It follows from Theorem 22 that $R_P/R \geq 1$.

In all our examples the intensity has independent jumps or is two-state markovian (which formally can also be represented as an independent jump intensity – see Remark 17). Recall from Theorem 34 that
$$R = \sup\{r \geq 0 \mid \phi(r) \leq 1\}$$
where
$$\phi(r) = E[e^{-rc\sigma + h(r)L\sigma}].$$
The parameters are chosen such that
$$\mu = E[Z] = 1 \quad \text{and} \quad \alpha = E[L\sigma]/E[\sigma] = 1.$$
Thus we have positive safety loading for $c > 1$.

CASE 1: Consider the case when $\sigma = s_0$ p-a.s. and let L be exponentially distributed with $E[L] = 1$. This is a special case of the model studied by Ammeter (1948) and discussed in Example 38. Then
$$\phi(r) = \frac{e^{-rcs_0}}{1 - h(r)s_0} \quad \text{when } h(r) < 1/s_0.$$

☐

We shall now give two examples of independent jump markovian intensities.

CASE 2: We consider first the case when $\eta(\ell) \equiv \eta$. Then L and σ are independent and $E[\sigma] = 1/\eta$. Recall from Example 26 that the distribution of L must have compact support, otherwise we have $\phi(r) = \infty$ for all $r > 0$. Let L be uniformly distributed on $[0, 2]$. It follows from Proposition 49 that
$$\phi(r) = -\frac{\eta}{2h(r)} \log\left(1 - \frac{2h(r)}{rc + \eta}\right) \quad \text{when } h(r) < \frac{rc + \eta}{2}.$$

☐

CASE 3: Now we consider a case when $P\{L > \ell\} > 0$ for all ℓ. Put
$$p_L(d\ell) = \begin{cases} 0 & \text{if } \ell < 1/2 \\ \frac{1}{2}\ell^{-2}\,d\ell & \text{if } \ell \geq 1/2 \end{cases} \quad \text{and} \quad \eta(\ell) = \eta\ell.$$
Then $E[\sigma] = 1/\eta$. It follows from Proposition 49 that
$$\phi(r) = \frac{\eta}{2rc}\log\left(1 + \frac{2rc}{\eta - h(r)}\right) \quad \text{when } h(r) < 2rc + \eta.$$

□

The last two examples illustrate two-state markovian intensities.

CASES 4 AND 5: In Case 4, which is meant to illustrate a "harmless" case, we put $\alpha_1 = 0.75$ and $\alpha_2 = 2$. In Case 5, which shall illustrate a "more dangerous" case, we put $\alpha_1 = 0.25$ and $\alpha_2 = 4$. In both cases we put $\eta_2 = 4\eta_1$. In order to make these illustrations comparable with the previous ones we put – see (38) – $E[\sigma] = \frac{1}{4}\left(\frac{1}{\eta_1} + \frac{1}{\eta_2}\right) = 5/(16\eta_1)$. For exponentially distributed claims these intensities were used as illustrations in Tables 1 and 2.

□

We consider the cases when Z is deterministic, exponentially distributed, and Γ-distributed with $\sigma_Z^2 = 100$, respectively. (We use the notation σ_Z^2 instead of just σ^2 in order to avoid confusion with the other σs appearing in this section.) Formally we can look upon Z as Γ-distributed with $\sigma_Z^2 = 0, 1,$ and 100. In these cases we have
$$h(r) = e^r - 1,$$
$$h(r) = \frac{r}{1-r} \quad \text{for } r < 1,$$
and
$$h(r) = (1 - 100r)^{-1/100} - 1 \quad \text{for } r < 1/100.$$

We refer to these three cases as "the deterministic case," "the exponential case," and "the Γ-case."

R_P is the positive solution of $h(r) = cr$, which in the exponential case yields $R_P = (c-1)/c$. In the deterministic case and the Γ-case – considered in Section 1.2 – R_P has to be solved numerically. For $c = 1.2$ we have

$$\begin{aligned} R_P &= 0.3542 &&\text{in the deterministic case;} \\ R_P &= 0.1667 &&\text{in the exponential case;} \\ R_P &= 0.003110 &&\text{in the }\Gamma\text{-case.} \end{aligned}$$

We have computed R_P/R for
$$E[\sigma] = 1000, 100, 10, 1, 0.1, \text{ and } 0.01$$
which are the same choices as in Tables 1 and 2. Those values are given in Tables 3 – 5.

4.6 Numerical illustrations

TABLE 3. Values of R_P/R for deterministic claims in the case $\alpha = \mu = 1$ and $c = 1.2$.

$E[\sigma]$	Case 1	Case 2	Case 3	Case 4	Case 5
1000	1130.112	577.258	836.689	409.070	3016.036
100	113.915	58.884	84.494	41.833	302.586
10	12.295	7.003	9.281	5.086	31.238
1	2.131	1.680	1.794	1.388	4.081
0.1	1.113	1.074	1.076	1.037	1.321
0.01	1.011	1.008	1.008	1.004	1.033

TABLE 4. Values of R_P/R for exponentially distributed claims in the case $\alpha = \mu = 1$ and $c = 1.2$.

$E[\sigma]$	Case 1	Case 2	Case 3	Case 4	Case 5
1000	532.241	272.234	394.131	192.972	1419.694
100	54.075	28.310	40.192	20.168	142.894
10	6.266	3.864	4.818	2.873	15.212
1	1.511	1.318	1.350	1.168	2.432
0.1	1.050	1.033	1.034	1.016	1.144
0.01	1.005	1.003	1.003	1.002	1.014

TABLE 5. Values of R_P/R for Γ-distributed claims in the case $\alpha = \mu = 1$ and $c = 1.2$.

$E[\sigma]$	Case 1	Case 2	Case 3	Case 4	Case 5
1000	10.811	6.177	8.167	4.495	27.438
100	1.932	1.554	1.645	1.308	3.620
10	1.088	1.058	1.059	1.028	1.255
1	1.009	1.006	1.006	1.003	1.025
0.1	1.001	1.001	1.001	1.000	1.003
0.01	1.000	1.000	1.000	1.000	1.000

In the discussion of Tables 1 and 2 we pointed out that small values of $E[\sigma]$ ought to mean that we are close to the Poisson case. Therefore, it is not surprising that R_P/R is close to one for small values of $E[\sigma]$. Furthermore, the tables indicate that the more harmless claim distribution the more influence the point process of occurrence of claims has. Although definite conclusions may not be drawn from these tables, we believe that the shown pattern holds rather generally.

One way to support that belief is to consider the diffusion approximation discussed in Section 1.2. Recall from (1.32) that it was natural to regard

$$R_{P_D} = \frac{2\rho\mu}{\mu^2 + \sigma_Z^2} = \frac{2(c - \alpha\mu)}{\alpha(\mu^2 + \sigma_Z^2)}$$

as the "diffusion approximation" of R_P. (Also recall the discussion about the diffusion approximation in the end of Section 1.2.) For $c = 1.2$ we have

$$\begin{aligned} R_{P_D} &= 0.4 & &\text{in the deterministic case;} \\ R_{P_D} &= 0.2 & &\text{in the exponential case;} \\ R_{P_D} &= 0.003960 & &\text{in the } \Gamma\text{-case.} \end{aligned}$$

Assume now that

$$\lim_{t \to \infty} \frac{\operatorname{Var}[\Lambda(t)]}{t} = \sigma_\Lambda^2$$

and that $\Lambda_n \xrightarrow{d} \sigma_\Lambda \cdot W$ as $n \to \infty$ where

$$\Lambda_n(t) = \frac{\Lambda(nt) - \alpha nt}{\sqrt{n}}.$$

These two assumptions are not too restrictive. It follows from Grandell (1976, p. 81) that $N_n \xrightarrow{d} \sqrt{\alpha + \sigma_\Lambda^2} \cdot W$ as $n \to \infty$ where

$$N_n(t) = \frac{N(nt) - \alpha nt}{\sqrt{n}}.$$

Then, see Grandell (1977, p. 47) for the result and Section 1.2 for the notation,

$$Y_n \xrightarrow{d} Y \quad \text{as} \quad n \to \infty,$$

where

$$Y_n(t) = \frac{c_n nt - \bar{S}(nt)}{\sqrt{n}} \quad \text{and} \quad Y(t) = \gamma\alpha\mu t - \sqrt{\mu^2\sigma_\Lambda^2 + \alpha(\mu^2 + \sigma_Z^2)} \cdot W(t)$$

if and only if

$$\rho_n \sqrt{n} = \frac{c_n - \alpha\mu}{\alpha\mu}\sqrt{n} \to \gamma \quad \text{as} \quad n \to \infty$$

similarly as in the Poisson case.

Thus it is natural to regard
$$R_D = \frac{2(c - \alpha\mu)}{\mu^2 \sigma_\Lambda^2 + \alpha(\mu^2 + \sigma_Z^2)}$$
as the "diffusion approximation" of R, and we are led to the approximation
$$\frac{R_P}{R} \approx \frac{R_{P_D}}{R_D} = \frac{\mu^2 \sigma_\Lambda^2 + \alpha(\mu^2 + \sigma_Z^2)}{\alpha(\mu^2 + \sigma_Z^2)}, \qquad (101)$$
which ought to hold for small values of $c - \alpha\mu$.

REMARK 57. In Section 1.2 we discussed the principal differences between approximations based on limit theorems and those more or less based on ad hoc reasoning. The approximation (101) must be regarded as based on ad hoc reasoning, although the difussion approximation is based on a limit theorem. The reason is partly that we have not proved that the ruin probabilities approximate each other, i.e., that $P\{\inf_{t\geq 0} Y_n(t) < -y\} \to P\{\inf_{t\geq 0} Y(t) < -y\}$, in the Cox case. However, even from such a proof it does not follow that $R_{P_D}/R_D \to R_P/R$ as $c \to \alpha\mu$. □

It follows from Asmussen (1987, p. 137) that
$$\sigma_\Lambda^2 = \frac{E[L^2\sigma^2] + \alpha^2 E[\sigma^2] - 2\alpha E[L\sigma^2]}{E[\sigma]} = \frac{E[(L-\alpha)^2\sigma^2]}{E[\sigma]}$$
for an independent jump intensity. In our cases we thus have
$$\frac{R_{P_D}}{R_D} = \frac{\sigma_\Lambda^2 + 1 + \sigma_Z^2}{1 + \sigma_Z^2}, \quad \text{where} \quad \sigma_\Lambda^2 = \frac{E[(L-1)^2\sigma^2]}{E[\sigma]}. \qquad (102)$$

By simple calculations we get
$$\sigma_\Lambda^2 = E[\sigma] \cdot \begin{cases} 1 & \text{in Case 1} \\ 2/3 & \text{in Cases 2 and 3} \\ 8/25 & \text{in Case 4} \\ 72/25 & \text{in Case 5} \end{cases}.$$

The simple approximation (102) holds reasonably well for $E[\sigma] \leq 10$ in the deterministic case, for $E[\sigma] \leq 100$ in the exponential case and for all $E[\sigma]$ in the Γ-case. Taking the poor accuracy of the diffusion approximation into account it holds, in our opinion, surprisingly well.

In Table 6 we consider the behavior of R_P/R for the "worst reasonable" values of $E[\sigma]$. The approximation R_{P_D}/R_D is indicated by "$c = 1$" and the claim distributions by the values of σ_Z^2.

In all cases, except in Case 4, R_P/R seems to increase or decrease to R_{P_D}/R_D. In Case 4, with exponentially distributed claims, R_P/R has a maximum 20.233 at $c = 1.236$. Generally R_P/R seems to be relatively insensitive to variation in c, and that is probably the reason why approximation (102) works reasonably – or surprisingly – well.

TABLE 6. Values of R_P/R (for $c > 1$) and R_{P_D}/R_D (indicated by $c = 1$) in the case $\alpha = \mu = 1$.

c	σ_Z^2	$E[\sigma]$	Case 1	Case 2	Case 3	Case 4	Case 5
1.3	0	10	12.914	6.665	10.054	5.269	31.606
1.2	0	10	12.295	7.003	9.281	5.086	31.238
1.1	0	10	11.657	7.337	8.468	4.773	30.648
1	0	10	11.000	7.667	7.667	4.200	29.800
1.3	1	100	55.484	25.546	42.748	20.052	140.625
1.2	1	100	54.075	28.310	40.192	20.168	142.894
1.1	1	100	52.584	31.220	37.404	19.246	144.426
1	1	100	51.000	34.333	34.333	17.000	145.000
1.3	100	1000	10.777	5.591	8.395	4.428	26.341
1.2	100	1000	10.811	6.177	8.167	4.495	27.438
1.1	100	1000	10.852	6.840	7.905	4.422	28.507
1	100	1000	10.901	7.601	7.601	4.168	29.515

Although definite conclusions may not be drawn from Tables 1 – 6, they do – in our opinion – support the conclusion that it might be fatal to ignore random fluctuations in the intensity process.

CHAPTER 5

Stationary models

Recall
$$\Psi(0) = \frac{\alpha\mu}{c} = \frac{1}{1+\rho} \qquad \text{when } c > \alpha\mu, \tag{I}$$

which was proved in Chapter 1 for the Poisson case. As pointed out (**I**) is an insensitivity result, since $\Psi(0)$ only depends on ρ and thus on F only through its mean μ. In Chapter 3 – see (3.39) – it was shown that (**I**) also holds for the *stationary* renewal model. Thus – in that case – (**I**) turned out to also depend on the inter-occurrence time distribution K^0 only through its mean $1/\alpha$. In Chapter 4 – see (4.19) – (**I**) was found to hold for a class of stationary Cox models. It was also shown that when N is a mixed Poisson process, i.e., a Cox process with $\lambda(t) \equiv \lambda$, then

$$\Psi(0) = \frac{\mu}{c} \int_0^{c/\mu} x \, dU(x) + \left(1 - U\left(\frac{c}{\mu}\right)\right), \tag{1}$$

where U is the distribution of the random variable λ.

In this chapter – based on Björk and Grandell (1985) – we consider the case when the occurrence of the claims is described by a stationary point process. It will be shown that (1), with a proper interpretation of U, holds for all stationary risk models.

REMARK 1. In the renewal model the calculation of ruin probabilities is strongly related to the calculation of waiting times in one-server queues, see Feller (1971, pp. 194 - 198).

Consider the $GI/G/1$ queue. This means that the customers arrive at a "server" according to an ordinary renewal process and that the service times are described by independent and identically distributed random variables. The arrivals and the services are also independent. Let, with the notation used in the renewal model, cS_{n-1} denote the arrival time of the nth customer – $S_0 \stackrel{\text{def}}{=} 0$ – and Z_n the service time of that customer. Assume for simplicity that K^0 is continuous.

Let η be the *traffic intensity*. The relation between the traffic intensity and the safety loading is given by

$$\eta \stackrel{\text{def}}{=} \frac{E[Z_n]}{E[c(S_n - S_{n-1})]} = \frac{\alpha\mu}{c} = \frac{1}{1+\rho} \quad \text{or} \quad \rho = \frac{1}{\eta} - 1.$$

Let W_n be the *waiting time* in the queue of the nth customer. Then we have

$$W_{n+1} = \begin{cases} W_n - c(S_n - S_{n-1}) + Z_n & \text{if } W_n - c(S_n - S_{n-1}) + Z_n \geq 0 \\ 0 & \text{if } W_n - c(S_n - S_{n-1}) + Z_n \leq 0 \end{cases}.$$

Put – see (3.2) and (3.4) –

$$X_k \stackrel{\text{def}}{=} Z_k - c(S_k - S_{k-1}) \quad \text{and} \quad Y_n = \sum_{k=1}^{n} X_k$$

and recall that

$$\Psi^0(u) = P\{\max_{n \geq 1} Y_n > u\}.$$

Thus we have $W_{n+1} = \max(0, W_n + X_n)$. Under the assumption that the server is idle at time 0, we get

$W_1 = 0,$

$W_2 = \max(0, X_1) = \max(0, Y_1),$

$W_3 = \max(W_2, X_2) = \max(0, X_2, X_1 + X_2) = \max(0, Y_2 - Y_1, Y_2),$

$W_4 = \max(W_3, X_3) = \max(0, X_3, X_2 + X_3, X_1 + X_2 + X_3)$
$\qquad = \max(0, Y_3 - Y_2, Y_3 - Y_1, Y_3),$

$\vdots$

$W_{n+1} = \cdots = \max(0, Y_n - Y_{n-1}, Y_n - Y_{n-2}, \ldots, Y_n - Y_1, Y_n).$

For $\eta < 1$ there exists a random variable W – called the *steady state* waiting time – such that $W_n \stackrel{\text{d}}{\to} W$ as $n \to \infty$. Since

$$(Y_n - Y_{n-1}, Y_n - Y_{n-2}, \ldots, Y_n - Y_1, Y_n) \stackrel{\text{d}}{=} (Y_1, Y_2, \ldots, Y_n), \qquad (2)$$

where $\stackrel{\text{d}}{=}$ means "equality in distribution," it follows that

$$\Psi^0(u) = P\{W > u\}.$$

The *virtual waiting time* $V(t)$ is the waiting time in the queue of a hypothetical customer arriving just after time t. When $\eta < 1$ there exists – since K^0 is continuous – a random variable V such that $V(t) \stackrel{\text{d}}{\to} V$ as $t \to \infty$ and we have

$$\Psi(u) = P\{V > u\}.$$

Note that $V(t) = 0$ if and only if the server is idle at time t.

Elementary books on queueing theory, and we choose Allen (1978) as an example, emphasize the $M/G/1$ queue – which means that the customers arrive according to a Poisson process – and the $M/M/1$ queue where the service times are exponentially distributed. In those cases W and V have the same distribution. These queues correspond to the Poisson case. From Allen (1978, p. 163) it follows that

$$P[W > u] = \eta e^{-\frac{(1-\eta)u}{\mu}} \quad \text{for the } M/M/1 \text{ queue}$$

which is the "queueing version" of (II) and from Allen (1978, p. 198) that

$$P\{V = 0\} = 1 - \eta \tag{3}$$

for the $M/G/1$ queue which is (I).

In more advanced treatises on classical queueing theory – the most well-known is probably Takács (1962) – it is shown that (Takács 1962, p. 142) (3) also holds for the $GI/G/1$ queue.

Like in risk theory it may be disputed if the $GI/G/1$ queue really is the relevant generalization of the $M/G/1$ queue. Franken et al. (1981) consider the much more general $G/G/1$ queue, where only certain stationarity properties of the arrivals and the services are assumed. It is, for example, not assumed that the arrivals and the services are independent. It follows from Franken et al. (1981, p. 108) that (3) still holds when the queueing system is ergodic. We will not go into details about the model and rely on the reader's intuition.

In this generality (2) does not necessarily hold and therefore the relation between ruin probabilities and waiting times is not quite problem-free. This "problem" seems, however, not too serious, since a time reversal may change distributions but not expectations and (I) only depends on expectations. In spite of this we will give a direct proof which generalizes the proof in the renewal case.

Björk and Grandell (1985, p. 149) gave an example which they claimed to be a "counter example" of that relation. Although that was not too well expressed – i.e., wrong – we shall consider the example. Let the claim sizes $Z_1, Z_2, Z_3, \ldots$ be independent and exponentially distributed with mean 1 and let the claims be located at $Z_1, Z_1 + Z_2, Z_1 + Z_2 + Z_3, \ldots$ Thus N is a Poisson process with $\alpha = 1$. In the queueing formulation (3) holds. Intuitively that is obvious, since the customers always arrive at an idle server. The nth customer arrives at time cS_{n-1}. That customer's service is completed at time $cS_{n-1} + Z_n$ while the next customer arrives at $cS_n = cS_{n-1} + cZ_n$. Thus the server is busy during $(cS_{n-1}, cS_{n-1} + Z_n)$ and idle during $(cS_{n-1} + Z_n, cS_n)$, i.e., the server is idle the proportion $(1-c)/c = 1 - \eta$ of the time. In the risk model formulation we have $X(t) \geq (c-1)t$ for all $t > 0$ and thus $\Psi(0) = 0$ for $c > 1$ which Björk and Grandell (1985, p. 149) regarded as a contradiction of (I). This is, however (in reality), no contradiction since the risk process $X(t)$ does not have stationary increments. In order to realize this, we consider an epoch $t_0 \gg 0$. Then – formally when $t_0 \to \infty$ – the time from t_0 to the next claim *is* exponentially distributed with mean 1. The time from the previous claim to t_0 is also exponentially distributed with mean 1 and the two durations are independent. Thus the risk process $X(t)$ gets stationary increments if the size of its first claim is changed to $Z_1 + \widetilde{Z}$ where $\widetilde{Z}$ is exponentially distributed with mean 1 and independent of all the Z_ks. Then ruin can occur only at the first claim and we have

$$\Psi(0) = P\{\widetilde{Z} > (c-1)Z_1\} = \int_0^\infty e^{-(c-1)z} e^{-z}\, dz = \int_0^\infty e^{-cz}\, dz = \frac{1}{c}$$

which is in agreement with (I).
□

As mentioned in Remark 1 we shall generalize the proof in the renewal case. In that case (I) followed from (3.37), which gave a relation between the ruin probabilities – or strictly speaking the non-ruin probabilities – in the ordinary and the stationary cases.

The natural question is now:

What is the correpondence to "the ordinary case" for a general stationary point process?

In order to answer that question we shall need some basic facts about stationary point processes. A good reference is Franken et al. (1981), upon which the survey is highly based.

STATIONARY POINT PROCESSES

We start by recalling some basic definitions given in the survey "Point processes and random measures" in Section 2.2.

Let $\mathcal{N}$ denote the set of integer or infinite valued Borel measures on $\mathbf{R} = (-\infty, \infty)$ and let $\mathcal{B}(\mathcal{N})$ denote the Borel algebra on $\mathcal{N}$. The elements in $\mathcal{N}$ are usually denoted by ν. A point process N is a measurable mapping from a probability space $(\Omega, \mathcal{F}, P)$ into $(\mathcal{N}, \mathcal{B}(\mathcal{N}))$. Its distribution is a probability measure Π on $(\mathcal{N}, \mathcal{B}(\mathcal{N}))$.

Put $\mathcal{N}_S = \{\nu \in \mathcal{N};\ \nu(t) - \nu(t-) = 0 \text{ or } 1\}$. Here we shall only consider simple point processes and therefore we omit the subscript S. With this convention any $\nu \in \mathcal{N}$ can be looked upon as a realization of a simple point process.

The shift operator $T_x : \mathcal{N} \to \mathcal{N}$ is defined by $(T_x \nu)\{A\} = \nu\{A + x\}$ for $A \in \mathcal{B}(\mathbf{R})$ and $x \in \mathbf{R}$ where $A + x = \{t \in \mathbf{R};\ t - x \in A\}$. We put $T_x B = \{\nu \in \mathcal{N};\ T_{-x}\nu \in B\}$ for any $B \in \mathcal{B}(\mathcal{N})$.

A point process is stationary if $\Pi\{T_x B\} = \Pi\{B\}$ for all $x \in \mathbf{R}$ and all $B \in \mathcal{B}(\mathcal{N})$. From now on Π is assumed to be the distribution of a stationary point process N with intensity $\alpha \in (0, \infty)$.

There always exists a random variable $\overline{N}$ with $E[\overline{N}] = \alpha$, called the *individual intensity*, such that $N(t)/t \to \overline{N}$ Π-a.s. as $t \to \infty$. Let $\mathcal{I}$ be the

σ-algebra of *invariant* sets $B \in \mathcal{B}(\mathcal{N})$, i.e., of sets B such that $B = T_x B$ for all $x \in \mathbf{R}$. N is *ergodic* if $\Pi\{B\} = 0$ or 1 for all $B \in \mathcal{I}$. Since $\{\nu \in \mathcal{N};\ \bar{\nu} \leq x\} \in \mathcal{I}$ for each x it follows that $\bar{N} = \alpha$ Π-a.s. if N is ergodic.

For any $B \in \mathcal{I}$ such that $0 < \Pi\{B\} < 1$ the conditional distribution $\Pi\{\cdot \mid B\}$ is stationary. Let $\nu_\emptyset$ denote the empty realization, i.e., $\nu_\emptyset\{A\} = 0$ for all $A \in \mathcal{B}(\mathbf{R})$. Thus Π has the unique representation

$$\Pi = p\Delta_\emptyset + (1-p)\Pi_\infty, \tag{4}$$

where $0 \leq p \leq 1$ and $\Delta_\emptyset$ and Π_∞ are probability measures on $(\mathcal{N}, \mathcal{B}(\mathcal{N}))$ such that $\Delta_\emptyset\{\{\nu_\emptyset\}\} = 1$ and $\Pi_\infty\{\{\nu_\emptyset\}\} = 0$. A realization of a stationary point process contains Π-a.s. zero or infinitely many points, and thus $\Pi_\infty\{\{\nu;\ \nu(\infty) = \infty\}\} = 1$. The distribution of an ergodic point process cannot be a non-trivial mixure of stationary distribution, and therefore $p = 0$ since $\alpha > 0$.

Now we shall consider "the correspondence to the ordinary case" in the question above. In the case of renewal processes we started with an ordinary renewal process and obtained a stationary renewal process by choosing the distribution of S_1 according to (3.1). If we start with a stationary renewal process the ordinary renewal process is obtained by conditioning upon the occurrence of a point at time 0. In terms of a stationary point process this means that we want to consider probabilities of the form

$$\Pi\{B \mid N\{\{0\}\} = 1\}.$$

The problem is thus to give such probabilities a precise meaning for a general stationary point process. Intuitively we consider an event B and successively shift the process so that its "points" fall at time 0. If this had been a statistical problem – and not the question of a probabilistic definition – we had probably considered the proportion of times when the shifted point process belonged to B. Instead, we now consider the ratio of certain related intensities.

Consider a set $B \in \mathcal{B}(\mathcal{N})$. Define the "$B$-thinned" process N^B by

$$N^B\{dx\} = 1_B(T_x N)N\{dx\}, \tag{5}$$

where – as usual – $1_B(N) = \begin{cases} 1 & \text{if } N \in B \\ 0 & \text{if } N \notin B \end{cases}$. This means that N^B consists of those points in N for which the shifted point process belongs to B. Obviously N^B is stationary.

Put $\mathcal{N}^0 = \{\nu \in \mathcal{N};\ \nu\{\{0\}\} = 1\}$ and note that $\mathcal{N}^0 \in \mathcal{B}(\mathcal{N})$ and that $N^{\mathcal{N}^0} = N$. Let $\alpha\{B\}$ be the intensity of N^B. It follows from Matthes et al. (1978, pp. 309 - 311) that $\alpha\{\cdot\}$ is a measure, i.e., σ-additive, on $(\mathcal{N}^0, \mathcal{B}(\mathcal{N}^0))$.

DEFINITION 2. Let N be a stationary point process with distribution Π. The distribution Π^0, defined by

$$\Pi^0\{B\} = \frac{\alpha\{B\}}{\alpha}, \qquad B \in \mathcal{B}(\mathcal{N}^0),$$

is called the *Palm distribution*.

$\Pi^0\{B\}$ is the strict definition of "$\Pi\{B \mid N\{\{0\}\} = 1\}$." For a precise interpretation of Π^0 as a conditional probability we refer to Franken et al. (1981, pp. 33 and 38).

Define the (random) shift operator θ by

$$\theta = T_{s_1(\nu)} \qquad \text{for } \nu \neq \nu_\emptyset$$

and recall that $s_1(\nu)$ is the epoch of the first point – or claim – after time zero.

It is sometimes convenient to extend Π^0 to $(\mathcal{N}, \mathcal{B}(\mathcal{N}))$ in the obvious way as follows: $\Pi^0\{B\} = \Pi^0\{B \cap \mathcal{N}^0\}$ for all $B \in \mathcal{B}(\mathcal{N})$.

The point process N^0 with distribution Π^0 is called the *Palm process*. N^0 is *not* stationary but

$$\Pi^0\{\theta B\} = \Pi^0\{B\} \qquad \text{for all } B \in \mathcal{B}(\mathcal{N}).$$

If $B \in \mathcal{I}$ and $\Pi\{B\} = 1$ it follows from (5) that $\Pi^0\{B\} = 1$. This means especially that $\overline{N}^0$ exists Π^0-almost surely. Let U be the distribution of $\overline{N}$ and U^0 the distribution of $\overline{N}^0$. Then we have, for $B = \{\nu; \nu \leq x\}$,

$$U^0(x) = \Pi^0\{B\} = \frac{\alpha\{B\}}{\alpha} = \frac{E[1_{[0,x]}(\overline{N})\overline{N}]}{\alpha} = \int_0^x y \, dU(y)/\alpha. \qquad (6)$$

For any non-negative $\mathcal{B}(\mathcal{N})$-measurable function f on $\mathcal{N}$ we have (Franken et al. 1981, pp. 26 - 27)

$$E_\infty[f(N)] = \alpha_\infty E^0\left[\int_0^{s_1(N)} f(T_t N) \, dt\right], \qquad (7)$$

where "N" just stands for a point process whose distribution is indicated by the notation of the expectation and where

$$\alpha_\infty \stackrel{\text{def}}{=} E_\infty[N(1)] = \alpha/(1-p).$$

For $f \equiv 1$ we get $E^0[S_1]$ ($= E^0[s_1(N)]$) $= 1/\alpha_\infty$ and thus (7) is an "inversion formula." (At least when $p = 0$, i.e., when $\Pi = \Pi_\infty$, $E^0[S_1] = 1/\alpha$ is the strict definition of "the mean duration between two successive claims.") Since, in general,

$$E[f(N)] = pf(\nu_\emptyset) + (1-p)E_\infty[f(N)]$$

we get

$$E[f(N)] = pf(\nu_\emptyset) + \alpha E^0\left[\int_0^{s_1(N)} f(T_t N) \, dt\right]. \qquad (8)$$

REMARK 3. In our attempt to give a heuristic motivation for Definition 2, we discussed the "proportion of times when the shifted point process belongs to B." In the ergodic case we have

$$\lim_{t\to\infty} \frac{N^B(t)}{N(t)} = \lim_{t\to\infty} \frac{N^B(t)}{t} \frac{t}{N(t)} = \frac{\alpha\{B\}}{\alpha} \quad \Pi\text{-a.s.}$$

and thus (Matthes et al. 1978, p. 339) we get the "correct" result.
□

If N is a stationary renewal process then (Matthes et al. 1978, p. 367) N^0 is the corresponding ordinary renewal process. Note that these renewal processes are defined on **R** and that N^0, as all Palm processes, has a point at 0. The superscript 0 is standard for Palm processes, and therefore we also used it in connection with renewal processes in order to indicate an ordinary renewal process.

EXAMPLE 4. We shall consider some examples of Palm processes. These examples will not be explicitly used, but they may support intuition.

It is often convenient to withdraw the point at 0 and to consider the *reduced* Palm process $N^!$ with distribution $\Pi^!$. Formally $N^!$ is defined by

$$N^!\{A\} = \begin{cases} N^0\{A\} - 1 & \text{if } 0 \in A \\ N^0\{A\} & \text{if } 0 \notin A \end{cases} \quad \text{for } A \in \mathcal{B}(\mathbf{R}).$$

If N is a Poisson process we have $\Pi = \Pi^!$ which is a characterization of the Poisson process. (This characterization also holds in the non-stationary case, although the Palm probability is somewhat differently defined.) Intuitively this means that knowledge of a point at 0 has no influence on the distribution of the rest of the process. This is quite natural, since the Poisson process is the only stationary point process with independent increments, and may be looked upon as a "Palm correspondence" to Theorem 2.11.

Assume that N is a Cox process with distribution Π_Λ given by, see (2.13), $\Pi_\Lambda = \int_{\mathcal{M}} \Pi_\mu \, \Pi\{d\mu\}$* and that Λ has the representation $\Lambda(t) = \int_0^t \lambda(s)\, ds$. It follows from Kummer and Matthes (1970, p. 1636) that $N^!$ is a Cox process with $\Pi^!_\Lambda = \int_{\mathcal{M}} \Pi_\mu \, \Pi^!\{d\mu\}$ where ($\mu'(0)$ exists Π-a.s.)

$$\Pi^!\{d\mu\} = \frac{\mu'(0)}{\alpha} \Pi\{d\mu\}. \tag{9}$$

(In Section 4.3 we considered "ordinary independent jump intensities" and "stationary independent jump intensities." Although the underlying ideas are related to Palm theory, the ordinary case is *not* the Palm process.)

---

* Note that Π and $\Pi^!$, in these comments about Cox processes, are distributions of random measures.

132   5 Stationary models

If N is a mixed Poisson process we have, of course, $\overline{N} = \lambda$ and (6) and (9) are in agreement.

☐ ☐

REMARK 1. CONTINUED. Assume that N is ergodic. Any customer who enters a queueing system also, hopefully, leaves it if $\eta < 1$. Then we ought to have

$$\underbrace{\alpha/c}_{\text{arrival intensity}} = \underbrace{P\{V > 0\}}_{\text{busy server}} \cdot \underbrace{1/\mu}_{\text{service intensity}}$$

or $\Psi(0) = P\{V > 0\} = \alpha\mu/c$ when $c > \alpha\mu$ which is (I).

In the non-ergodic case, α ought to be replaced by $\overline{N}$. Obviously V and $\overline{N}$ are dependent. Then we ought to have

$$\underbrace{\overline{N}/c}_{\text{arrival intensity}} = \underbrace{P\{V > 0 \mid \overline{N}\}}_{\text{busy server}} \cdot \underbrace{1/\mu}_{\text{service intensity}}$$

and "thus" $P\{V > 0 \mid \overline{N}\} = \overline{N}\mu/c$ when $c > \overline{N}\mu$. "Thus" $\Psi(0) = E[P\{V > 0 \mid \overline{N}\}]$ which is (1) if U is the distribution of $\overline{N}$.

Certainly this reasoning shall not be taken too seriously, but it may serve as an indication of the kind of results to be expected.

☐

Now we consider the risk process. Let N be the restriction of a point process on $\mathbf{R}$ to $\mathbf{R}_+$. As in the survey Π is the stationary distribution and Π^0 the Palm distribution. Recall that $N(0) = 0$ for all point processes on $\mathbf{R}_+$ and therefore we do not need to separate between the Palm process and the reduced Palm process.

Let $\Psi(u, \nu)$ be the ruin probability when the claims are located according to the realization ν of N. Thus

$$\Psi(u) = E[\Psi(u, N)] \quad \text{where } E \text{ is with respect to } \Pi.$$

Put

$$\Psi^0(u) = E^0[\Psi(u, N)] \quad \text{where } E^0 \text{ is with respect to } \Pi^0.$$

The following lemma may be of some independent interest.

LEMMA 5. *For any stationary risk model with* $0 < \alpha < \infty$ *we have*

$$\Psi(0) = \frac{\alpha\mu}{c}(1 - \Psi^0(\infty)) + \Psi(\infty).$$

PROOF: Put $\Phi(u) = 1 - \Psi(u)$, $\Phi^0(u) = 1 - \Psi^0(u)$, and $\Phi(u, \nu) = 1 - \Psi(u, \nu)$. For fixed ν and t, $0 \leq t < s_1(\nu)$, we have – "standing" at time t –

$$\Phi(u, T_t\nu) = \int_0^{u+c(s_1(\nu)-t)} \Phi(u + c(s_1(\nu) - t) - z, \theta\nu) \, dF(z) \qquad (10)$$

by a slight variant of the "renewal" argument used in Sections 1.1 and 3.2.

For $t = 0$ we get, denoting $s_1(N)$ by S_1,

$$\Phi^0(u) = E^0\left[\int_0^{u+cS_1} \Phi(u + cS_1 - z, \theta N)\, dF(z)\right]. \tag{11}$$

Using (8) with $f(\nu) = \Phi(u, \nu)$ we get, since $\Phi(u, \nu_\emptyset) = 1$,

$$\Phi(u) = p + \alpha E^0\left[\int_0^{S_1} \Phi(u, T_t N)\, dt\right]$$

and by (10)

$$\Phi(u) = p + \alpha E^0\left[\int_0^{S_1}\int_0^{u+cv} \Phi(u + cv - z, \theta N)\, dF(z)\, dv\right]. \tag{12}$$

The change of variables $x = u + cv$ leads to

$$\Phi(u) = p + \frac{\alpha}{c} E^0\left[\int_u^{u+cS_1}\int_0^x \Phi(x - z, \theta N)\, dF(z)\, dx\right] \tag{13}$$

which is almost the same as (3.35). From Lemma 3.8 applied to

$$\varphi(x, \nu) = \int_0^x \Phi(x - z, \nu)\, dF(z)$$

it follows that differentiation of both sides of (13) is allowed, and we get

$$\Phi'(u) = \frac{\alpha}{c}\Phi^0(u) - \frac{\alpha}{c}\int_0^u \Phi^0(u - z)\, dF(z) \tag{14}$$

which is exactly (3.36). Thus we have

$$\Phi(\infty) = \Phi(0) + \frac{\alpha\mu}{c}\Phi^0(\infty)$$

which is (3.38) and the lemma is proved. ∎

Now we can formulate the main result.

THEOREM 6. *For any stationary risk model $0 < \alpha < \infty$ we have*

$$\Psi(0) = \frac{\mu}{c}\int_0^{c/\mu} x\, dU(x) + \left(1 - U\left(\frac{c}{\mu}\right)\right),$$

where U is the distribution of the individual intensity $\overline{N}$ of the underlying stationary point process N.

PROOF: Assume first that U has no mass point at c/μ, i.e., that $U(c/\mu) - U(c/\mu -) = 0$.

Since the claim costs are independent it follows, by the strong law of large numbers, that

$$\lim_{t\to\infty} X(t)/t = c - \overline{N}\mu \quad P\text{-a.s.} \tag{15}$$

Note that (15) is trivially true if $N = \nu_\emptyset$ and that $0 < \lim_{t\to\infty} N(t) < \infty$ is impossible Π-a.s. From this (for $\overline{\nu} > c/\mu$) and from the arguments between

(1.5) and (**I**) in Chapter 1 (for $\bar{\nu} < c/\mu$) it follows that

$$\Psi(\infty,\nu) = \lim_{u\to\infty} \Psi(u,\nu) = \begin{cases} 0 & \text{for any } \nu \text{ with } \bar{\nu} < c/\mu \\ 1 & \text{for any } \nu \text{ with } \bar{\nu} > c/\mu \end{cases}.$$

For $\bar{\nu} = c/\mu$ nothing can be said in general, but this case only occurs with probability 0.

Thus
$$\Psi(\infty) = 1 - U(c/\mu).$$

Since we have only used the existence of $\bar{N}$, and not the stationarity of N, the above reasoning also holds in the Palm case. Since, see (6), $dU^0(x) = x dU(x)/\alpha$ it follows that

$$1 - \Psi^0(\infty) = U^0(c/\mu) = \int_0^{c/\mu} y dU(y)/\alpha$$

and the result in the theorem holds.

Assume next that $U(c/\mu) - U(c/\mu-) = 1$ which implies that $\alpha = c/\mu$. For any $c' > c$ the proved part of the theorem can be applied, and we have $1 \geq \Psi(\infty) \geq \alpha\mu/c' = c/c'$. Since this holds for any $c' > c$ the result in the theorem holds in this case also.

Consider finally the case $0 < U(c/\mu) - U(c/\mu-) < 1$ and put $q = U(c/\mu) - U(c/\mu-)$, $B = \{\nu;\ \bar{\nu} = c/\mu\}$, and $B^* = \{\nu;\ \bar{\nu} \neq c/\mu\}$. Since B and $B^* \in \mathcal{I}$ the conditional distributions $\Pi^B \stackrel{\text{def}}{=} \Pi\{\cdot \mid B\}$ and $\Pi^{B^*} \stackrel{\text{def}}{=} \Pi\{\cdot \mid B^*\}$ are stationary and we have $\Pi = q\Pi^B + (1-q)\Pi^{B^*}$. Thus we have, with obvious notation, $U(x) = qU^B(x) + (1-q)U^{B^*}(x)$ where

$$U^B(x) = \begin{cases} 0 & \text{if } x < c/\mu \\ 1 & \text{if } x \geq c/\mu \end{cases}$$

and

$$U^{B^*}(x) = \begin{cases} (1-q)^{-1}U(x) & \text{if } x < c/\mu \\ (1-q)^{-1}(U(x)-q) & \text{if } x \geq c/\mu \end{cases}.$$

Put $\Psi^B(u) = \int_{\mathcal{N}} \Psi(u,\nu)\Pi^B\{d\nu\}$ and $\Psi^{B^*}(u) = \int_{\mathcal{N}} \Psi(u,\nu)\Pi^{B^*}\{d\nu\}$ and note that the proved parts can be applied to Ψ^B and Ψ^{B^*}. Thus we have, since $\alpha^B = c/\mu$,

$$\Psi(0) = q\Psi^B(0) + (1-q)\Psi^{B^*}(0)$$

$$= q \cdot 1 + (1-q)\left(\frac{\mu}{c}\int_0^{c/\mu} x\, dU^{B^*}(x) + \left(1 - U^{B^*}\left(\frac{c}{\mu}\right)\right)\right)$$

$$= q + \frac{\mu}{c}\int_0^{c/\mu -} x\, dU(x) + 1 - U\left(\frac{c}{\mu}\right) = \frac{\mu}{c}\int_0^{c/\mu} x\, dU(x) + 1 - U\left(\frac{c}{\mu}\right)$$

and the theorem is proved. ∎

We may note that if $c > \alpha\mu$, which is the interesting case, and if $U(c/\mu) = 1$, then (**I**) holds although N is not ergodic. This means that (**I**) holds when the risk business is "profitable," i.e, when $c > \bar{N}\mu$, with probability one.

APPENDIX
Finite time ruin probabilities

Up to now we have only considered the probability of ruin within infinite time, i.e., the probability that the risk business *ever* becomes negative.

Let a time t be given and let – as usual – T_u denote the time of ruin. The *finite time ruin probability* $\Psi(u,t)$ is defined by

$$\Psi(u,t) = P\{T_u \leq t\} = P\{u + X(s) < 0 \text{ for some } s \in (0,t]\}.$$

From a practical point of view, $\Psi(u,t)$, where t is related to the planning horizon of the company, may perhaps sometimes be regarded as more interesting than $\Psi(u)$. Most insurance managers will closely follow the development of the risk business and increase the premium if the risk business behaves disquietingly bad. Also an orthodox probabilist will probably act in the same way, since he – despite a wish to keep his job – will believe that the underlying model is wrong. In this connection the planning horizon may be thought of as the sum of the following:

the time until the risk business is found to behave "bad";

the time until the manager reacts;

the time until a decision of a premium increase takes effect.

It may therefore, in non-life insurance, be natural to regard t equal to four or five years as reasonable.

Depending on the branch and the company it may be reasonable to consider α – when the time unit is years – of the orders 10^3 to 10^5. Just to have some value in mind, we regard 50000 as a reasonable value of $\alpha \cdot t$.

The intention of this appendix is to give some indication on when the infinite time ruin probability is also relevant for the finite time case. Intuitively one may expect that ruin – if ever – occurs as follows:

after a short time if u is small and ρ is large;

after a long time if u is large and ρ is small.

More precisely, there exists – at least in some cases – a value y_0, such that
$$\Psi(u,t) \sim \Psi(u) \text{ when } t > uy_0$$
while
$$\Psi(u,t) << \Psi(u) \text{ when } t < uy_0$$
for large values of u. Otherwise expressed, this means that *either* $T_u = \infty$, i.e., no ruin, *or* $T_u \approx uy_0$, i.e., ruin.

Our interpretation is that $\Psi(u)$ is most relevant when the planning horizon is longer than uy_0.

This does, however, not imply that we regard $\Psi(u)$ as irrelevant when the planning horizon is shorter than uy_0. It seems quite natural to look beyond the first possibility to adjust the premium when the initial reserve u is determined.

To start with we consider the classical model in some detail.

A.1 The classical model

Recall from (1.19) that
$$\Psi(u,t) \leq e^{-ru} \sup_{0 \leq s \leq t} e^{s(\alpha h(r)-rc)} = e^{-ru} \max(1, e^{t(\alpha h(r)-rc)}). \tag{1}$$

Obviously we can always choose $r = R$, but it might be possible – at least sometimes – to choose a better, i.e., a larger, value of r.

Put $t = yu$. Then (1) yields
$$\Psi(u, yu) \leq \max(e^{-ru}, e^{-u(r-y\alpha h(r)+yrc)}) = e^{-u\min(r, r-y\alpha h(r)+yrc)}$$

and it seems natural to define the *"time-dependent"* Lundberg exponent R_y by
$$R_y = \sup_{r \geq 0} \min(r, r - y\alpha h(r) + yrc) = \sup_{r \geq R}(r - y\alpha h(r) + yrc)$$

and we have the *"time-dependent"* Lundberg inequality (Gerber 1973, p. 208)
$$\Psi(u, yu) \leq e^{-R_y u}. \tag{2}$$

Put
$$f(r) = r - y\alpha h(r) + yrc$$

and note that $f(R) = R$, $f(r) < r$ for $r > R$ and that $f(r)$ is concave. Thus we have, since $R_y \geq R$,
$$R_y \stackrel{=}{>} R \text{ as } f'(R) \stackrel{\leq}{>} 0.$$

Since $f'(R) = 1 - y\alpha h'(R) + yc$ it follows that
$$R_y \begin{array}{c} = \\ > \end{array} R \quad \text{as} \quad y \begin{array}{c} \geq \\ < \end{array} \frac{1}{\alpha h'(R) - c}.$$
The value $y_0 \stackrel{\text{def}}{=} \frac{1}{\alpha h'(R)-c}$ is called the *critical value*. For $y < y_0$ we have $R_y = f(r_y)$ where r_y is the solution of $f'(r) = 0$.

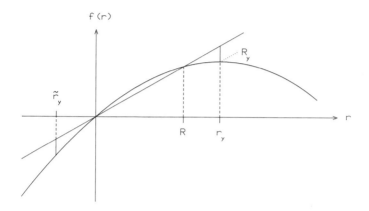

FIGURE 1. Illustration of notation when $y < y_0$.

It follows from Arfwedson (1955, pp. 58 and 78) that
$$\Psi(u, yu) \sim \begin{cases} \frac{C_y}{\sqrt{u}} e^{-R_y u} & \text{if } y < y_0 \\ \frac{C}{2} e^{-Ru} & \text{if } y = y_0 \\ C e^{-Ru} & \text{if } y > y_0 \end{cases} \quad \text{as } u \to \infty \qquad (3)$$
and thus R_y is the "right" exponent. The constant C_y is given by
$$C_y = -\frac{r_y - \tilde{r}_y}{r_y \tilde{r}_y \sqrt{2\pi y \alpha h''(r_y)}},$$
provided the claim distribution is non-arithmetic, where $\tilde{r}_y$ is the negative solution of
$$f(r) - r = R_y - r_y,$$
and C is the constant in the Cramér-Lundberg approximation (**III**).

Segerdahl (1955, p. 34) has shown that
$$\Psi(u, t) \sim \mathcal{N}\left(\frac{t - uy_0}{\sqrt{uv_0}}\right) C e^{-Ru} \qquad (4)$$
as $u, t \to \infty$, and $\frac{t-uy_0}{\sqrt{u}}$ is bounded where $\mathcal{N}(x)$ denotes the standard normal distribution function, i.e.,
$$\mathcal{N}(x) = \int_{-\infty}^{x} \frac{1}{\sqrt{2\pi}} e^{-\frac{z^2}{2}} \, dz$$

and
$$v_0 = \frac{\alpha h''(R)}{(\alpha h'(R) - c)^3}. \tag{5}$$

(Our use of $\mathcal{N}$ instead of the more usual Φ is because we want to reserve the notation Φ for the non-ruin probability.)

It is pointed out by Segerdahl (1955, p. 35) that (4) may be improved if uy_0 is replaced by

$$y_0(u) = uy_0 + \frac{v_0}{y_0} - \frac{1}{\alpha\mu R\rho} - \frac{\sigma^2 + \mu^2}{2\alpha\mu^2\rho^2} \tag{6}$$

and uv_0 by a similar ($uv_0 +$ const.) expression. This is especially true for "reasonable" values of u and "dangerous" claim distributions.

It follows from (4) that T_u conditioned upon $\{T_u < \infty\}$ is asymptotically normally distributed with mean uy_0 and a variance uv_0. This implies – for large values of u – that ruin, if it occurs, occurs in the time interval $(uy_0 - 2\sqrt{uv_0}, uy_0 + 2\sqrt{uv_0})$ with approximately the probability 0.95.

Our conclusion is that the study of ruin probabilities within infinite time is also relevant for the the finite time case when $t > y_0 u$ and u is large.

EXAMPLE 1. EXPONENTIALLY DISTRIBUTED CLAIMS. We have, see Example 1.7,

$$R = \frac{\rho}{\mu(1+\rho)}, \quad \Psi(u) = \frac{1}{1+\rho}e^{-Ru}, \quad \text{and} \quad h'(R) = \mu(1+\rho)^2$$

and thus
$$y_0 = \frac{1}{\alpha\mu\rho(1+\rho)} \quad \text{and} \quad v_0 = \frac{2}{\alpha^2\mu\rho^3}.$$

For $y < y_0$ we then get, compare Arfwedson (1955, p. 82),

$$\mu r_y = 1 - \sqrt{\frac{\alpha\mu y}{1 + \alpha\mu y(1+\rho)}}, \quad R_y = \alpha y \left(\frac{\mu r_y}{1 - \mu r_y}\right)^2, \quad \text{and}$$

$$\mu \tilde{r}_y = 1 - \frac{1}{(1+\rho)(1 - \mu r_y)}$$

and thus
$$C_y = -\frac{(r_y - \tilde{r}_y)(1 - \mu r_y)^{3/2}}{2\mu r_y \tilde{r}_y \sqrt{y\alpha\pi}}.$$

It may be natural to choose u by specifying $\Psi(u)$. Then we get

$$\alpha \cdot uy_0 = -\frac{\log[(1+\rho)\Psi(u)]}{\rho^2} \quad \text{and}$$

$$\alpha^2 \cdot uv_0 = -\frac{2(1+\rho)\log[(1+\rho)\Psi(u)]}{\rho^4}. \tag{7}$$

In Table 1 we give u, $\alpha \cdot uy_0$, and $\alpha \cdot \sqrt{uv_0}$ for some values of ρ and $\Psi(u)$.

TABLE 1. Values of u, $\alpha \cdot uy_0$, and $\alpha \cdot \sqrt{uv_0}$ for exponentially distributed claims when $\mu = 1$.

ρ	$\Psi(u)$	u	$\alpha \cdot uy_0$	$\alpha \cdot \sqrt{uv_0}$
1%	0.01	464	45952	30467
1%	0.05	302	29858	24559
1%	0.10	232	22926	21520
5%	0.01	96	1823	1237
5%	0.05	62	1179	995
5%	0.10	47	902	870
10%	0.01	50	451	315
10%	0.05	32	290	253
10%	0.10	24	221	220
20%	0.01	27	111	81
20%	0.05	17	70	65
20%	0.10	13	53	56

It is seen that $\alpha \cdot uy_0$ and $\alpha \cdot \sqrt{uv_0}$ are of the same order for the values chosen in Table 1. This indicates that u has to be much larger than in Table 1 before is it reasonable to say that $T_u \approx uy_0$ when ruin occurs. The "asymptotic coefficient of variation" is

$$\frac{\alpha \cdot \sqrt{uv_0}}{\alpha \cdot uy_0} = \frac{\sqrt{2\mu} \cdot (1+\rho)}{\sqrt{u\rho}},$$

which indicates that u has to be "very" large.

The correction term in (6) has only little influence since

$$\alpha \cdot y_0(u) = \frac{u}{\mu\rho(1+\rho)} + \frac{1}{\rho}.$$

If, for example, $\rho = 0.05$, $\alpha = 1$, and $\mu = 1$ we have (Wikstad 1971, p. 149)

$$\Psi(100, 1000) = 0.0019 \quad \text{and} \quad \Psi(100) = 0.0081.$$

For this choice we get $uy_0 = 1905$ and $\sqrt{uv_0} = 1265$ and thus

$$\mathcal{N}\left(\frac{1000 - 1905}{1265}\right) \Psi(100) \approx 0.24 \cdot 0.0081 = 0.0019$$

which indicates that (4) holds with good accuracy in this case. For a more detailed discussion we refer to Asmussen (1984, pp. 37 - 42).

Further, we have, see (3),

$$\begin{aligned} y &= 10, \\ r_y &= 0.067495, \\ R_y &= 0.052389, \\ \tilde{r}_y &= -0.021315, \\ C_y &= 4.958834, \end{aligned}$$

and thus

$$\Psi_A(100, 1000) = 0.0026,$$

where Ψ_A is the "Arfwedson approximation" given by (3). We have also computed $\Psi_A(10, 100) = 0.9287$ which shall be compared with $\Psi(10, 100) = 0.3464$.

□

EXAMPLE 2. WIENER PROCESS WITH DRIFT. As indicated in Section 1.1 the martingale approach also applies to a Wiener process with drift. Let W be a standard Wiener process and consider

$$X(t) = \beta t + \delta W(t), \qquad \beta > 0.$$

Then

$$E[e^{-rX(t)}] = e^{t(-\beta r + \delta^2 r^2/2)}$$

which formally means that $\alpha h(r) - rc$ is replaced by $-\beta r + \delta^2 r^2/2$. Thus

$$R = \frac{2\beta}{\delta^2}, \qquad y_0 = \frac{1}{\delta^2 R - \beta} = \frac{1}{\beta}, \quad \text{and} \quad v_0 = \frac{\delta^2}{\beta^3}. \qquad (8)$$

Further, for $y < y_0$, we have $f(r) = r + y\beta r - y\delta^2 r^2/2$ and thus

$$r_y = \frac{1 + \beta y}{y\delta^2}, \qquad R_y = \frac{(1+\beta y)^2}{2y\delta^2}, \quad \text{and} \quad \tilde{r}_y = -\frac{1 - \beta y}{y\delta^2}.$$

In this case we have

$$\Psi(u,t) = 1 - \mathcal{N}\left(\frac{u + \beta t}{\delta\sqrt{t}}\right) + \mathcal{N}\left(\frac{-u + \beta t}{\delta\sqrt{t}}\right) e^{-Ru} \qquad (9)$$

or

$$\Psi(u, yu) = 1 - \mathcal{N}\left(\frac{\beta y + 1}{\delta\sqrt{y}}\sqrt{u}\right) + \mathcal{N}\left(\frac{\beta y - 1}{\delta\sqrt{y}}\sqrt{u}\right) e^{-Ru}. \qquad (10)$$

Using the approximation, see Cramér (1955, p. 38),

$$\left.\begin{array}{r}1 - \mathcal{N}(x) \\ \mathcal{N}(-x)\end{array}\right\} \sim \frac{1}{x\sqrt{2\pi}} e^{-\frac{x^2}{2}} \quad \text{as } x \to \infty$$

we get for $y < y_0$:

$$\Psi(u, yu) \sim \frac{\delta\sqrt{y}}{\sqrt{2\pi u}(1+\beta y)} e^{-\frac{(1+\beta y)^2}{2y\delta^2}u} + e^{-\frac{2\beta}{\delta^2}u} \cdot \frac{\delta\sqrt{y}}{\sqrt{2\pi u}(1-\beta y)} e^{-\frac{(1-\beta y)^2}{2y\delta^2}u}$$

$$= \frac{1}{\sqrt{2\pi y\delta^2 \cdot u}} \left(\frac{1}{r_y} - \frac{1}{\tilde{r}_y} \right) e^{-R_y u} = -\frac{r_y - \tilde{r}_y}{r_y \tilde{r}_y \sqrt{2\pi y\delta^2}} \frac{1}{\sqrt{u}} e^{-R_y u},$$

for $y = y_0$:

$$\Psi(u, yu) \sim \frac{\delta}{2\sqrt{2\pi\beta u}} e^{-Ru} + \tfrac{1}{2} e^{-Ru} \sim \frac{1}{2} e^{-Ru},$$

and for $y > y_0$:

$$\Psi(u, yu) \sim \frac{\delta\sqrt{y}}{\sqrt{2\pi u}(1+\beta y)} e^{-\frac{(1+\beta y)^2}{2y\delta^2}u} + 1 \cdot e^{-\frac{2\beta}{\delta^2}u} \sim e^{-Ru}.$$

Thus it is seen that (3) holds in this case.

Assume now that $t = y_0 u + \epsilon\sqrt{u}$. Then

$$\Psi(u,t) \sim 1 - \mathcal{N}\left(\frac{2u + \beta\epsilon\sqrt{u}}{\delta\sqrt{y_0 u + \epsilon\sqrt{u}}} \right) + \mathcal{N}\left(\frac{\beta\epsilon\sqrt{u}}{\delta\sqrt{y_0 u + \epsilon\sqrt{u}}} \right) e^{-Ru}$$

$$\sim \frac{\text{const.}}{\sqrt{u}} e^{-\frac{2u(u+\beta\epsilon\sqrt{u})}{\delta^2\beta^2(u+\beta\epsilon\sqrt{u})}} + \mathcal{N}\left(\frac{\epsilon}{\delta/\beta^{3/2}} \right) e^{-Ru}$$

$$= \frac{\text{const.}}{\sqrt{u}} e^{-Ru} + \mathcal{N}\left(\frac{\epsilon}{\delta/\beta^{3/2}} \right) e^{-Ru} \sim \mathcal{N}\left(\frac{\epsilon}{\delta/\beta^{3/2}} \right) e^{-Ru}.$$

Thus (4) also holds.
□

From Table 1 it seems highly relevant to consider the infinite time ruin probability when $\rho \geq 5\%$. Certainly no general conclusions may be drawn from this simple case.

Consider now any claim distribution and assume that R and $\Psi(u_0)$ are known for some u_0 so large that (**III**) is a good approximation. Then

$$\Psi(u_0) \approx Ce^{-Ru_0} \quad \text{or} \quad C \approx \Psi(u_0)e^{Ru_0}.$$

Specifying $\Psi(u)$ we thus get $\Psi(u) \approx \Psi(u_0)e^{-R(u-u_0)}$ or

$$u \approx u_0 + \frac{\log[\Psi(u_0)/\Psi(u)]}{R}. \tag{11}$$

Next we observe, compare (**III**) and (1.26), that $C = \alpha y_0 \cdot \rho\mu$ or

$$\alpha y_0 = \frac{C}{\rho\mu} = \frac{\Psi(u_0)}{\rho\mu} e^{Ru_0} \tag{12}$$

and thus

$$\alpha \cdot u y_0 = \frac{\Psi(u_0)}{\rho\mu} \left(u_0 + \frac{\log[\Psi(u_0)/\Psi(u)]}{R} \right) e^{Ru_0}. \tag{13}$$

Naturally it is desirable to choose u_0 such that $\Psi(u_0) \approx \Psi(u)$. If $u_0 = u$ (13) is reduced to

$$\alpha \cdot u y_0 = \frac{u}{\rho\mu} \Psi(u) e^{Ru} \leq \frac{u}{\rho\mu}, \tag{14}$$

where the inequality follows from (**IV**).

It may be natural to exploit the simple De Vylder approximation, discussed in Section 1.1, which worked so well in the infinite time case. Recall that the idea was to replace the risk process X with a risk process $\tilde{X}$ having exponentially distributed claims and parameters

$$\tilde{\mu} = \frac{\zeta_3}{3\zeta_2}, \quad \tilde{\rho} = \frac{2\zeta_1\zeta_3}{3\zeta_2^2}\rho, \quad \text{and} \quad \tilde{\alpha} = \frac{9\zeta_2^3}{2\zeta_3^2}\alpha,$$

where $\zeta_k = E[Z_j^k]$. Applying (6) to $\tilde{X}$ we get

$$\tilde{\alpha} \cdot \widetilde{uy_0} = -\frac{\log[(1+\tilde{\rho})\Psi(u)]}{\tilde{\rho}^2}$$

and, if $\widetilde{uy_0}$ is a good approximation of uy_0,

$$\alpha \cdot uy_0 \approx \alpha \cdot \widetilde{uy_0} = \frac{\alpha}{\tilde{\alpha}} \cdot \tilde{\alpha} \cdot \widetilde{uy_0} = -\frac{\log[(1+\tilde{\rho})\Psi(u)]}{\frac{\tilde{\alpha}}{\alpha}\tilde{\rho}^2}$$

$$= -\frac{\zeta_2 \log[(1 + \frac{2\zeta_1\zeta_3}{3\zeta_2^2}\rho)\Psi(u)]}{2\zeta_1^2 \rho^2}. \tag{15}$$

EXAMPLE 3. Γ-DISTRIBUTED CLAIMS. We consider the case with $\rho = 10\%$ and where the claims are Γ-distributed with $\mu = 1$ and $\sigma^2 = 100$, which was discussed in Section 1.2. From Table 1.1 it follows that $\Psi(1200) = 0.10834$. Since $R = 0.0017450$ it follows from (11) and (13) that $\Psi(u) = 0.1$ corresponds to

$$u \approx 1246 \quad \text{and} \quad \alpha \cdot uy_0 \approx 10957.$$

Applying the De Vylder approximation (15) we get

$$\alpha \cdot \widetilde{uy_0} = 10999,$$

which is almost perfect.

□

EXAMPLE 4. MIXED EXPONENTIALLY DISTRIBUTED CLAIMS. Consider now the claim distribution (1.35)

$$F(z) = 1 - 0.0039793 e^{-0.014631 z} -$$

$$0.1078392 e^{-0.190206 z} - 0.8881815 e^{-5.514588 z} \quad \text{for } z \geq 0$$

discussed in Section 1.2. Using Tables 1.2 and 1.3 we get from (12), (10), and (14) the values of u, $\alpha \cdot uy_0$ and $\alpha \cdot \widetilde{uy_0}$ given in Table 2. We have, however, used more accurate R-values, than those given in Table 1.2. For $\rho = 5\%$ and 10% we used $u_0 = 1000$ and otherwise $u_0 = 100$.

TABLE 2. Values of u, $\alpha \cdot uy_0$, $\alpha \cdot \widetilde{uy_0}$, and $\alpha \cdot \sqrt{uv_0}$ for mixed exponentially distributed claims when $\Psi(u) = 0.1$.

ρ	u	$\alpha \cdot uy_0$	$\alpha \cdot \widetilde{uy_0}$	$\alpha \cdot \sqrt{uv_0}$
5%	1068	18703	18778	19214
10%	567	4386	4447	4946
15%	398	1830	1878	2253
20%	312	967	1006	1293
25%	258	582	615	841
30%	222	381	408	592
100%	72	16	21	54

As for exponentially distributed claims we see that $\alpha \cdot uy_0$ and $\alpha \cdot \sqrt{uv_0}$ are of the same order for the values chosen.

As mentioned in Section 1.2, this claim distribution has been considered by Wikstad (1971). The values in Table 3 are taken from Wikstad (1971, p. 151).

TABLE 3. Mixed exponentially distributed claims.

u	ρ	$\alpha \cdot uy_0$	$\Psi(u, 10)$	$\Psi(u, 100)$	$\Psi(u, 1000)$	$\Psi(u)$
100	5%	1751	0.0094	0.0896	0.4115	0.7144
100	10%	773	0.0094	0.0863	0.3618	0.5393
100	15%	460	0.0093	0.0833	0.3186	0.4247
100	20%	310	0.0093	0.0804	0.2813	0.3455
100	25%	225	0.0092	0.0777	0.2493	0.2886
100	30%	172	0.0092	0.0751	0.2219	0.2461
100	100%	23	0.0087	0.0497	0.0723	0.0724
1000	5%	17505	0.0000	0.0000	0.0004	0.1149
1000	10%	7734	0.0000	0.0000	0.0003	0.0210
1000	15%	4600	0.0000	0.0000	0.0002	0.0054
1000	20%	3105	0.0000	0.0000	0.0002	0.0018

Like in Example 1 we consider $\rho = 0.05$. For $u = 100$ we have $\alpha \cdot \sqrt{uv_0} = 5878$ and $\alpha \cdot y_0(u) = 3000$. First we note that correction term in (6) has – as to be expected – high influence. Further, we have

$$\mathcal{N}\left(\frac{1000 - 1751}{5878}\right) \Psi(100) \approx 0.449 \cdot 0.7144 = 0.321$$

which shall be compared with $\Psi(100, 1000) = 0.4115$. This indicates that (4) does not hold with good accuracy in this case. Especially we note $1000 < \alpha \cdot 100 y_0$ while $\Psi(100, 1000) > \frac{1}{2}\Psi(100)$. Our conclusion is that $u = 100$ is "too small" in this case.

For $u = 1000$ we have $\alpha \cdot \sqrt{uv_0} = 18589$. Then

$$\mathcal{N}\left(\frac{1000 - 17505}{18589}\right)\Psi(1000) = \mathcal{N}(-0.89)\Psi(1000) \approx 0.187 \cdot 0.1149 = 0.021$$

which is of a different order than $\Psi(1000, 1000) = 0.0004$. We do admit that this total lack of accuracy is very surprising, especially since $\mathcal{N}(-0.89)$ is not a "tail value." The reader's — and certainly our — first reaction is probably that there is a computional error. The following crude estimates do, however, indicate that this is not the case. We have, for $R = 0.002$,

$$h'(R) \approx \zeta_1 + R \cdot \zeta_2 + \frac{R^2}{2} \cdot \zeta_3 \approx 1.102 \quad \text{and} \quad h''(R) \approx \zeta_2 + R \cdot \zeta_3 \approx 58.63$$

and thus, for $u = 1000$,

$$\alpha \cdot uy_0 \approx \frac{1}{1.102 - 1.05} \approx 19300 \quad \text{and} \quad \alpha \cdot \sqrt{uv_0} \approx 20500$$

which indicates that the lack of accuracy is not due to a computional error.

For these two choices of u and t we have

y	$= 10,$		y	$= 1,$
r_y	$= 0.002651,$		r_y	$= 0.007864,$
R_y	$= 0.002167,$		R_y	$= 0.005578,$
$\tilde{r}_y$	$= -0.000742,$		$\tilde{r}_y$	$= -0.011365,$
C_y	$= 25.303321,$		C_y	$= 4.387626$

and thus

$$\Psi_A(100, 1000) = 2.037, \qquad \Psi_A(1000, 1000) = 0.000525.$$

Obviously $u = 100$ is also "too small" in this case while $u = 1000$ leads to a reasonable — although not very good — approximation. □

It is, as always, difficult to draw conclusions from a few numerical illustrations. It seems, however, as the statement that "either $T_u = \infty$ or $T_u \approx uy_0$" require very large values of u in order to be "true." This, in turn, makes uy_0 large. In order to claim that $\Psi(u, t) \approx \Psi(u)$ it is, of course, enough that $P\{T_u > t \mid T_u < \infty\}$ is small, which ought to be the case more generally as soon as $t >> uy_0$.

Our way to first choose ρ and then determine u by specifying $\Psi(u)$ is natural from a "risk theoretic" point of view. For a dangerous risk distribution we thus get a large value of u and consequently a large value of uy_0. We could have argued in an "opposite" way, i.e., to first choose u and then determine ρ by specifying $\Psi(u)$. Certainly the approximations (3) and (4) had not worked, but $P\{T_u > t \mid T_u < \infty\}$ had probably become small. In "the theoretical" practice it seems natural to allow ρ, u, and $\Psi(u)$ to be large for a dangerous risk distribution. In "the real" practice ρ and u are probably chosen more by economical – than by risk theoretic – considerations. Some practical working actuaries consider $u = \alpha \cdot \mu$ to be reasonable. Then it follows from (14) that $uy_0 \leq 1/\rho$. If $\rho = 20\%$, which also is regarded as practically reasonable, we have $uy_0 \leq 5$ years $\approx$ the planning horizon.

A.2 Renewal models

Let N be an ordinary renewal process and let S_n denote the epoch of the nth claim. The Laplace transform of the inter-occurrence time distribution is denoted by $\hat{k}^0$. Recall from Chapter 3 that R is the positive solution of $(h(r) + 1)\hat{k}^0(cr) = 1$.

The asymptotic expressions (4) and (3) have been generalized by von Bahr (1974) and Höglund (1990), respectively. Their analysis is based on the two-dimensional random walk

$$\{(X(S_n), S_n); \ n = 0, 1, 2, \ldots\}.$$

Put, in order to simplify notation,

$$H_k = \begin{cases} h(R) + 1 & \text{if } k = 0 \\ h^{(k)}(R) & \text{if } k > 0 \end{cases} \quad \text{and} \quad K_k = (\hat{k}^0)^{(k)}(cR),$$

where (k) denotes the kth derivative. Note that $H_0 K_0 = 1$. It follows from von Bahr (1974, pp. 201 and 203) that (4) holds with

$$y_0 = \frac{-H_0 K_1}{H_1 K_0 + cH_0 K_1} \quad \text{and} \quad v_0 = \frac{H_0 H_2 K_1^2 + H_1^2 K_0 K_2 - 2H_1^2 K_1^2}{(H_1 K_0 + cH_0 K_1)^3}. \quad (16)$$

REMARK 5. In the Poisson case we have, see Remark 3.4, $K_1 = -\frac{1}{\alpha} K_0^2$ and $K_2 = \frac{2}{\alpha^2} K_0^3$. Thus (16) reduces to

$$y_0 = \frac{-\frac{1}{\alpha} K_0}{H_1 K_0 + \frac{c}{\alpha} K_0} = \frac{1}{\alpha H_1 - c}$$

and

$$v_0 = \frac{\frac{1}{\alpha^2} H_2 K_0^3}{(H_1 K_0 + cH_0 K_1)^3} = \frac{\alpha H_2}{(\alpha H_1 - c)^3}.$$

□

In the classical case we proved a "time-dependent" Lundberg inequality by a simple extension of the martingale proof of the usual Lundberg inequality.

The analysis in Chapter 3 was based on the random walk
$$\{- X(S_n);\ n = 0, 1, 2, \ldots\}$$
and the number N_u of the claim causing ruin. Since we are interested in $T_u = S_{N_u}$ – and not in N_u itself – the approach in Chapter 3 will hardly be of any help.

The problem with generalizing the "continuous time" analysis in the classical case in a straightforward manner is that $X(t)$ is – in general – not markovian. It is, however, easy to obtain markovity by introducing a supplementary process. Following Dassios and Embrechts (1989, pp. 198 - 200) we consider the vector process
$$(X,\ V) = \{(X(t),\ V(t));\ t \geq 0\},$$
where $V(t)$ is the time elapsed since the last claim, and the filtration **F** given by $\mathcal{F}_t = \mathcal{F}_t^X \vee \mathcal{F}_t^V$. In the ordinary case there is a claim at the origin, i.e., $V(0) = 0$, and then $\mathcal{F}_t = \mathcal{F}_t^X$. In order to reduce the number of parentheses in the formulas below, we – sometimes – use the notations X_t and V_t. The **F**-compensator $\hat{\Lambda}$ of the underlying renewal process N – see Definition 2.13 – is given by
$$\hat{\Lambda}(t) = \int_0^t l(V_s)\,ds, \qquad \text{where } l(t) = \frac{k^0(t)}{1 - K^0(t)}.$$

For any fixed $r < r_\infty$, i.e., such that $h(r) < \infty$, we look for a positive **F**-martingale $M_u(t)$ of the form
$$M_u(t) = e^{-\vartheta(t)} q(V_t) e^{-r(u+X_t)},$$
where ϑ and q – which, of course, may depend on r – are differentiable. Without loss of generality we further assume that $\vartheta(0) = 0$ and $q(0) = 1$. Note that $q(t) > 0$.

Using the markovity of $(X,\ V)$ we get*
$$E^{\mathcal{F}_t}[M_u(t + \Delta)]$$
$$= (1 - l(V_t)\Delta) \cdot e^{-\vartheta(t+\Delta)} q(V_t + \Delta) e^{-r(u+c\Delta+X_t)} +$$
$$l(V_t) \cdot \Delta e^{-\vartheta(t)} q(0) e^{-r(u+X_t)} (h(r) + 1) + o(\Delta)$$
$$= M_u(t) + e^{-r(u+X_t)+\vartheta(t)} \cdot \Delta \cdot$$
$$\cdot \left[- \vartheta'(t) q(V_t) + q'(V_t) - c r q(V_t) + l(V_t)((h(r) + 1) - q(V_t))\right] + o(\Delta).$$

Thus M_u is an **F**-martingale if
$$q'(v) - (\vartheta'(t) + cr) q(v) - l(v) q(v) + l(v)(h(r) + 1) = 0.$$
Since the function $q(v)$ does not depend on t, we must have
$$\vartheta(t) = \theta \cdot t$$

---
* Compare "Davis' observation" discussed in Section A.3.

and thus q is a solution of
$$q'(v) - (\theta + cr)q(v) - l(v)q(v) + l(v)(h(r) + 1) = 0.$$
Multiplication with $e^{-(\theta+cr)v + \log(1-K^0(v))} = (1 - K^0(v))e^{-(\theta+cr)v}$ yields
$$\frac{d}{dv}\left(q(v)(1 - K^0(v))e^{-(\theta+cr)v}\right) = -(h(r) + 1)k^0(v)e^{-(\theta+cr)v}$$
which has the general solution
$$q(v)(1 - K^0(v))e^{-(\theta+cr)v} = (h(r) + 1)\int_v^\infty e^{-(\theta+cr)s} k^0(s)\, ds + C$$
or
$$q(v) = \frac{e^{(\theta+cr)v}}{1 - K^0(v)}\left((h(r) + 1)\int_v^\infty e^{-(\theta+cr)s} k^0(s)\, ds + C\right).$$
For $v = 0$ we get
$$1 = (h(r) + 1)\hat{k}^0(\theta + cr) + C$$
which, since $q(v) > 0$, implies that $0 \leq C < 1$.

Consider now $M_u(S_1)$. Since $V(S_1) = 0$ we get
$$M_u(S_1) = e^{\theta S_1} \cdot \left((h(r) + 1)\hat{k}^0(\theta + cr) + C\right) \cdot e^{-r(u+cS_1-Z_1)}$$
$$= e^{-ru-(\theta+cr)S_1+rZ_1}$$
and thus
$$E[M_u(S_1)] = e^{-ru}(h(r) + 1)\hat{k}^0(\theta + cr) = M_u(0)(h(r) + 1)\hat{k}^0(\theta + cr).$$
It is natural to require – see Theorem 1.14 – that $E[M_u(S_1)] = M_u(0)$, which holds only if $C = 0$.

Thus (Dassios and Embrechts 1989, p. 199)
$$M_u(t) = \frac{e^{-r(u+X_t)-\theta t+(\theta+cr)V_t}}{1 - K^0(V_t)}(h(r) + 1)\int_{V_t}^\infty e^{-(\theta+cr)s} k^0(s)\, ds$$
is an **F**-martingale provided θ and r are related through
$$(h(r) + 1)\hat{k}^0(\theta + cr) = 1.$$

The aim of Dassios and Embrechts (1989) is to show that many important risk processes belong to the class of so-called piecewise-deterministic Markov processes which was introduced by Davis (1984). The process (X, V) belongs to that class. In their derivation the conclusion $C = 0$ follows from a certain regularity condition. Our requirement $E[M_u(S_1)] = M_u(0)$ can be looked upon as an indirect regularity condition.*

---

* In the stationary case V_0 is a random variable with density $\alpha(1 - K^0(v))$. Then
$$E[M_u(0)] = E[q(V_0)]e^{-ru} > ce^{-ru}\alpha \int_0^\infty e^{(\theta+cr)v}\, dv = \infty,$$
since $\hat{k}^0(\theta + cr) = (1-c)/(h(r)+1) < 1$ implies $\theta + cr > 0$, and $C = 0$ follows without any "extra" regularity conditions.

Consider $\theta = \theta(r)$ as a function of r and note that $\theta(R) = 0$. The functions $h(r) + 1$ and $\hat{k}^0(r)$ are logarithmically convex, i.e., the functions χ and κ given by

$$\chi(r) = \log(h(r) + 1) \quad \text{and} \quad \kappa(r) = \log(\hat{k}^0(r))$$

are convex. This follows since $\chi''(r) = \text{Var}[U] > 0$, where U is a random variable with distribution $dF_U(z) = e^{rz} dF(z)/(h(r)+1)$. In the same way $\kappa'' > 0$ follows. Thus we have

$$\chi(r) + \kappa(\theta(r) + cr) = 0.$$

By derivation we get

$$\chi'(r) + \kappa'(\theta(r) + cr)(\theta'(r) + c) = 0 \quad \text{or} \quad \theta'(r) = -\frac{\chi'(r) + \kappa'(\theta(r) + cr)}{\kappa'(\theta(r) + cr)}$$

and thus

$$\theta'(0) = \alpha\mu - c < 0 \quad \text{and} \quad \theta'(R) = -\frac{\chi'(R) + \kappa'(cR)}{\kappa'(cR)} = -\frac{H_1 K_0 + c H_0 K_1}{H_0 K_1}.$$

A further derivation yields

$$\chi''(r) + \kappa''(\theta(r) + cr)(\theta'(r) + c)^2 + \kappa'(\theta(r) + cr)\theta''(r) = 0.$$

Since $\chi''(r) > 0$, $\kappa''(\theta(r) + cr) > 0$, $(\theta'(r) + c)^2 > 0$ while $\kappa'(\theta(r) + cr) < 0$ it follows that $\theta''(r) > 0$, i.e., $\theta(r)$ is convex.

Note that, since $T_u = S_{N_u}$,

$$M_u(T_u) = e^{-r(u+X(T_u)) - \theta(r)T_u} \quad \text{on } \{T_u < \infty\}.$$

As in the classical case we now get

$$\Psi^0(u, yu) \leq \frac{e^{-ru}}{E[M_u(T_u) \mid T_u \leq yu]}$$

$$\leq \frac{e^{-ru}}{E[e^{-T_u g(r)} \mid T_u \leq yu]} \leq e^{-u \min(r, r - y\theta(r))}.$$

Put $f(r) = r - y\theta(r)$ and $R_y = \sup_{r \geq 0} \min(r, r - y\theta(r))$ and it follows that

$$R_y \genfrac{}{}{0pt}{}{=}{>} R \quad \text{as} \quad y \genfrac{}{}{0pt}{}{\geq}{<} \frac{1}{\theta'(R)} = \frac{-H_0 K_1}{H_1 K_0 + c H_0 K_1} = y_0.$$

For $y < y_0$ we have

$$R_y = f(r_y) \quad \text{where } r_y \text{ is the solution of } f'(r) = 0.$$

EXAMPLE 6. MIXED EXPONENTIALLY DISTRIBUTED INTER-OCCURRENCE TIMES. Consider the inter-occurrence times distribution

$$K^0(t) = 1 - p_1 e^{-\theta_1 t} - p_2 e^{-\theta_2 t} \quad \text{for } t \geq 0 \tag{17}$$

with Laplace transform

$$\hat{k}^0(v) = \frac{p_1 \theta_1}{\theta_1 + v} + \frac{p_2 \theta_2}{\theta_2 + v},$$

where $\theta_2 \geq \theta_1 > 0$, $p_k \geq 0$, and $p_1 + p_2 = 1$. Obviously

$$\alpha = \frac{1}{\frac{p_1}{\theta_1} + \frac{p_2}{\theta_2}} = \frac{\theta_1 \theta_2}{p_1 \theta_2 + p_2 \theta_1}.$$

Since

$$K_k = (-1)^k \left(\frac{k! p_1 \theta_1}{(\theta_1 + cR)^{k+1}} + \frac{k! p_2 \theta_2}{(\theta_2 + cR)^{k+1}} \right),$$

y_0 and v_0 follows from (16).

Now we consider y_0 in some detail. It follows from (16) that

$$y_0 = \frac{1}{-\frac{H_1 K_0}{H_0 K_1} - c} \quad \text{or} \quad \frac{1}{y_0} = -\frac{H_1 K_0}{H_0 K_1} - c. \tag{18}$$

Since

$$\hat{k}^0(v) = \frac{\theta_1 \theta_2 + (p_1 \theta_1 + p_2 \theta_2) v}{(\theta_1 + v)(\theta_2 + v)}, \tag{19}$$

we get

$$K_0 = \frac{\theta_1 \theta_2 + (p_1 \theta_1 + p_2 \theta_2) cR}{(\theta_1 + cR)(\theta_2 + cR)}$$

and, by differentiation of (19),

$$K_1 = \frac{(p_1 \theta_1 + p_2 \theta_2) - K_0(2cR + \theta_1 + \theta_2)}{(\theta_1 + cR)(\theta_2 + cR)}$$

and thus

$$\frac{1}{y_0} = \frac{H_1[\theta_1 \theta_2 + (p_1 \theta_1 + p_2 \theta_2) cR]}{H_0[p_1 \theta_1 + p_2 \theta_2] - [2cR + \theta_1 + \theta_2]} - c.$$

Recall from Theorem 2.38 and Example 2.37 that N is a Cox process. In Example 2.37 we considered a Cox process N whose intensity process $\lambda(t)$ was a two-state Markov processes with $\alpha_1 = 0$. It was shown that N is a renewal process with K^0 given by (17). It follows from (19) and the form of $\hat{k}^0$ given in Example 2.37 that

$$p_1 \theta_1 + p_2 \theta_2 = \alpha_2,$$
$$\theta_1 \theta_2 = \alpha_2 \eta_1,$$
$$\theta_1 + \theta_2 = \alpha_2 + \eta_1 + \eta_2.$$

Using these relations we get

$$\frac{1}{y_0} = \frac{-H_1[\alpha_2 \eta_1 + \alpha_2 cR]}{H_0 \alpha_2 - (2cR + \alpha_2 + \eta_1 + \eta_2)} - c$$

$$= \frac{-h'(R) \alpha_2 [cR + \eta_1]}{h(R) \alpha_2 - (2cR + \eta_1 + \eta_2)} - c. \tag{20}$$

Like in Section 3.3 we now consider the special choice

$$p_1 = 0.25, \quad p_2 = 0.75, \quad \theta_1 = 0.4, \quad \text{and} \quad \theta_2 = 2$$

which corresponds to
$$\alpha_2 = 1.6, \quad \eta_1 = 0.5, \quad \text{and} \quad \eta_2 = 0.3.$$
Recall from (3.10) that
$$\Psi^0(u) = (1 - \mu R)e^{-Ru}$$
when the claims are exponentially distributed. (The superscript 0 refers to the "ordinary case.")

In Table 4 we give u, $\alpha \cdot uy_0$, and $\alpha \cdot \sqrt{uv_0}$ for the same values of ρ and $\Psi(u)$ as given in Table 1.

TABLE 4. Values of u, $\alpha \cdot uy_0$, and $\alpha \cdot \sqrt{uv_0}$ for exponentially distributed claims when $\mu = 1$ and mixed exponentially distributed inter-occurrence times.

ρ	$\Psi^0(u)$	u	$\alpha \cdot uy_0$	$\alpha \cdot \sqrt{uv_0}$
1%	0.01	811	80512	53276
1%	0.05	527	52340	42955
1%	0.10	405	40206	37648
5%	0.01	166	3205	2154
5%	0.05	107	2078	1734
5%	0.10	82	1593	1518
10%	0.01	85	798	546
10%	0.05	55	515	439
10%	0.10	42	394	383
20%	0.01	45	198	140
20%	0.05	29	127	112
20%	0.10	22	97	98

Roughly speaking, Table 4 and Table 1 give a very similar impression. As we did in Example 1 we consider approximation (4). For $\rho = 0.05$, $\alpha = 1$, and $\mu = 1$ we have (Wikstad 1971, p. 150)
$$\Psi^0(100, 1000) = 0.0196 \quad \text{and} \quad \Psi^0(100) = 0.0614.$$
For this choice we get $uy_0 = 1935$ and $\sqrt{uv_0} = 1673$ and thus
$$\mathcal{N}\left(\frac{1000 - 1935}{1673}\right)\Psi^0(100) \approx 0.29 \cdot 0.0614 = 0.0177$$
which indicates that (4) holds with reasonable accuracy in this case.

Consider now the claim distribution discussed in Example 4. Some illustrations of this combination of inter-occurrence time and claim distribution were given in Section 3.3.

The values in Table 5, which shall be compared to Table 3, are taken from Wikstad (1971, p. 151).

TABLE 5. Mixed exponentially distributed claims and mixed exponentially distributed inter-occurrence times.

u	ρ	$\alpha \cdot uy_0$	$\Psi^0(u,10)$	$\Psi^0(u,100)$	$\Psi^0(u,1000)$	$\Psi^0(u)$
100	5%	1760	0.0103	0.0932	0.4209	0.7231
100	10%	780	0.0103	0.0898	0.3710	0.5502
100	15%	465	0.0103	0.0867	0.3274	0.4356
100	20%	315	0.0102	0.0837	0.2897	0.3557
100	25%	229	0.0102	0.0809	0.2572	0.2978
100	30%	174	0.0101	0.0782	0.2293	0.2544
100	100%	23	0.0096	0.0519	0.0754	0.0754
1000	5%	17596	0.0000	0.0000	0.0005	0.1225
1000	10%	7804	0.0000	0.0000	0.0003	0.0232
1000	15%	4654	0.0000	0.0000	0.0002	0.0061
1000	20%	3148	0.0000	0.0000	0.0002	0.0020

Naturally we can give a table corresponding to Table 2. Since Tables 3 and 5 are almost identical, this would be of very restricted interest. □

EXAMPLE 7. LIFE ANNUITY INSURANCE. Like in Example 1.8 we now consider the case $c < 0$. (We do not assume $F(0) = 1$, as was done in Example 1.8.) Ruin does not need to occur at claim epochs and consequently the martingale approach used in Chapter 3 does not work. Note that – in that approach – only $c \geq 0$, and not $F(0) = 0$, is of importance. The martingale approach used here does, however, also work in this case.

Assume that $(h(r) + 1)\hat{k}^0(cr) = 1$, where – in this example – $h(r) = \int_{-\infty}^{\infty} e^{rz} \, dF(z) - 1$, has a solution $R > 0$. Since $\theta(R) = 0$ we have, for $r = R$,

$$M_u(t) = \frac{e^{-R(u+X_t)+cRV_t}}{1 - K^0(V_t)} (h(R) + 1) \int_{V_t}^{\infty} e^{-cRs} k^0(s) \, ds.$$

Noting that $c < 0$ we get by our "usual" arguments

$$\Psi^0(u) \leq \frac{M_u(0)}{E[M_u(T_u) \mid T_u < \infty]}$$

$$= \frac{e^{-Ru} \cdot (h(R) + 1)\hat{k}^0(cR)}{E\left[\frac{e^{-R(u+X(T_u))+cRV(T_u)}}{1-K^0(V(T_u))} (h(R) + 1) \int_{V(T_u)}^{\infty} e^{-cRs} k^0(s) \, ds \,\Big|\, T_u < \infty\right]}$$

$$= \frac{e^{-Ru}}{E\left[\frac{e^{-R(u+X(T_u))+cRV(T_u)}}{1-K^0(V(T_u))}(h(R)+1)\int_{V(T_u)}^{\infty} e^{-cRs}k^0(s)\,ds \mid T_u < \infty\right]}$$

$$\leq \frac{e^{-Ru}}{E\left[\frac{e^{cRV(T_u)}}{1-K^0(V(T_u))}(h(R)+1)\int_{V(T_u)}^{\infty} e^{-cRs}k^0(s)\,ds \mid T_u < \infty\right]}$$

$$\leq \frac{e^{-Ru}}{E\left[\frac{1}{1-K^0(V(T_u))}(h(R)+1)\int_{V(T_u)}^{\infty} k^0(s)\,ds \mid T_u < \infty\right]}$$

$$= \frac{e^{-Ru}}{h(R)+1} = \hat{k}^0(cR)e^{-Ru},$$

a result due to Thorin (1971, p. 31). Since $\hat{k}^0(cR) > 1$ this inequality is slightly weaker than in the classical case. Although $\hat{k}^0(cR)$ can be replaced by the constant one (Thorin 1971, p. 141) when $F(0) = 1$ and the interoccurrence times are mixed exponentially distributed, this can (Thorin 1971, pp. 139 - 140) – in general – *not* be done.

Consider now the stationary case. Then V_0 is a random variable with density $\alpha(1 - K^0(v))$, and we have

$$\Psi(u) \leq \frac{E[M_u(0)]}{E[M_u(T_u) \mid T_u < \infty]} \leq e^{-Ru} E\left[\frac{e^{cRV_0}}{1-K^0(V_0)}\int_{V_0}^{\infty} e^{-cRs}k^0(s)\,ds\right]$$

$$= e^{-Ru}\alpha \int_0^{\infty} e^{cRv}\int_v^{\infty} e^{-cRs}k^0(s)\,ds\,dv$$

$$= \frac{\alpha(1-\hat{k}^0(cR))}{cR}e^{-Ru} = \frac{\alpha h(R)}{cR}\frac{e^{-Ru}}{h(R)+1}$$

From (3.8) and (3.42) it is seen that there is the same relation between the ordinary case and the stationary case as when $c \geq 0$.
□

A.3 Cox models

Very little seems to be known in the Cox case, except when the intensity process $\lambda(t)$ is an M-state Markov process. Put, as in Remark 4.56,

$$\mathbf{H}(r) = h(r)d(\boldsymbol{\alpha}) - rc\mathbf{I} + \mathbf{H},$$

where $\boldsymbol{\alpha}$ is the "state vector" and $\mathbf{H}$ the intensity matrix. Recall that 0 is an eigenvalue of $\mathbf{H}(R)$ with maximal real part.

Let $\kappa(r)$ be the eigenvalue of $\mathbf{H}(r)$ with maximal real part. Asmussen (1989, p. 80) has shown that $\kappa(r)$ is convex and further $\kappa(0) = \kappa(R) = 0$.

Thus R is the positive solution of $\kappa(r) = 0$. As mentioned in the introduction to Chapter 4, Asmussen (1989) considers the Cramér-Lundberg approximation in this case. More precisely, Asmussen (1989, p. 92) shows that
$$\lim_{u \to \infty} e^{Ru} \Psi_i(u) = C_i,$$
where $\Psi_i(u)$ denotes the ruin probability when $\lambda(0) = \alpha_i$.

Further Asmussen (1989, p. 94) shows that (4) holds with
$$y_0 = \frac{1}{\kappa'(R)} \quad \text{and} \quad v_0 = \frac{\kappa''(R)}{\kappa'(R)^3} \qquad (21)$$
for all initial values of λ.

Now we restrict ourselves to the case $M = 2$, i.e., the case where the intensity is a two-state Markov process. Then $\mathbf{H} = \begin{pmatrix} -\eta_1 & \eta_1 \\ 0.4a_2 & -\eta_2 \end{pmatrix}$. In this case
$$\kappa(r) = \frac{1}{2}\Big(h(r)(\alpha_1 + \alpha_2) - 2cr - (\eta_1 + \eta_2) + \sqrt{[h(r)(\alpha_1 - \alpha_2) - (\eta_1 - \eta_2)]^2 + 4\eta_1\eta_2} \,\Big)$$
and
$$\kappa'(r)$$
$$= \frac{1}{2}\left(h'(r)(\alpha_1 + \alpha_2) - 2c + \frac{[h(r)(\alpha_1 - \alpha_2) - (\eta_1 - \eta_2)]h'(r)(\alpha_1 - \alpha_2)}{\sqrt{[h(r)(\alpha_1 - \alpha_2) - (\eta_1 - \eta_2)]^2 + 4\eta_1\eta_2}} \right)$$
and thus we get
$$\kappa'(R)$$
$$= \frac{1}{2}\left(h'(R)(\alpha_1 + \alpha_2) - 2c - \frac{[h(R)(\alpha_1 - \alpha_2) - (\eta_1 - \eta_2)]h'(R)(\alpha_1 - \alpha_2)}{h(R)(\alpha_1 + \alpha_2) - 2cR - (\eta_1 + \eta_2)} \right)$$
$$= \frac{2h(R)h'(R)\alpha_1\alpha_2 - h'(R)[cR(\alpha_1 + \alpha_2) + \alpha_1\eta_2 + \alpha_2\eta_1]}{h(R)(\alpha_1 + \alpha_2) - 2cR - (\eta_1 + \eta_2)} - c.$$

REMARK 8. For $\alpha_1 = \alpha_2 = \alpha$, i.e., in the Poisson case, we have
$$\kappa'(r) = \frac{1}{2}(h'(r)2\alpha - 2c - 0) = h'(r)\alpha - c \quad \text{and} \quad \kappa''(r) = h''(r)\alpha,$$
i.e.,
$$y_0 = \frac{1}{h'(r)\alpha - c} \quad \text{and} \quad v_0 = \frac{\alpha h''(R)}{(\alpha h'(R) - c)^3}.$$

For $\alpha_1 = 0$, i.e., in the renewal case, $\kappa'(R)$ reduces to
$$\kappa'(R) = \frac{-h'(R)\alpha_2[cR + \eta_1]}{h(R)\alpha_2 - 2cR - (\eta_1 + \eta_2)} - c$$
which equals (20).
□

EXAMPLE 9. Consider the case when the claims are exponentially distributed with $\mu = 1$. As in Section 4.1 – see Tables 4.1 and 4.2 – we specify a model by α_1 and $E[\sigma]$. Then

$$\eta_1 = 0.0003125, \quad \eta_2 = 0.00125 \quad \text{for } E[\sigma] = 1000;$$
$$\eta_1 = 0.03125, \quad \eta_2 = 0.125 \quad \text{for } E[\sigma] = 10;$$
$$\eta_1 = 3.125, \quad \eta_2 = 12.5 \quad \text{for } E[\sigma] = 0.1.$$

The use of $E[\sigma]$ as a characteristic came from the representation of a two-state Markov process as a Markov process with independent jumps.

TABLE 6. Two-state Markov intensity and exponentially distributed claims in the case $\alpha = \mu = 1$.

α_1	α_2	$E[\sigma]$	y_0 for $\rho = 5\%$	y_0 for $\rho = 10\%$	y_0 for $\rho = 15\%$	y_0 for $\rho = 20\%$	y_0 for $\rho = 25\%$
0.00	5	1000	19.3006	9.3471	6.0566	4.4296	3.4664
0.25	4	1000	19.0889	9.1682	5.9062	4.3042	3.3633
0.50	3	1000	18.6955	8.8600	5.6687	4.1266	3.2374
0.75	2	1000	17.7308	8.2580	5.3424	4.0232	3.3253
1.00	1	1000	19.0477	9.0909	5.7971	4.1667	3.2000
0.00	5	10	19.2817	9.3275	6.0363	4.4085	3.4445
0.25	4	10	19.0816	9.1547	5.8873	4.2805	3.3352
0.50	3	10	18.7682	8.8932	5.6707	4.1028	3.1915
0.75	2	10	18.4236	8.5970	5.4254	3.9071	3.0417
1.00	1	10	19.0477	9.0909	5.7971	4.1667	3.2000
0.00	5	0.1	19.0580	9.1012	5.8074	4.1769	3.2102
0.25	4	0.1	19.0477	9.0911	5.7974	4.1670	3.2004
0.50	3	0.1	19.0458	9.0892	5.7954	4.1650	3.1983
0.75	2	0.1	19.0471	9.0904	5.7966	4.1662	3.1995
1.00	1	0.1	19.0477	9.0909	5.7971	4.1667	3.2000

The most striking impression of Table 6 is certainly that y_0 essentially only depends on ρ.
□

Let us now consider the case when $\lambda(t)$ is a markovian jump process. The discussion here will be closely related to the discussion in Section 4.5.2 and notation used there will not be redefined here. Let us, however, recall that in Section 4.5.1 – where we applied our general method – we considered the filtration **F** given by $\mathcal{F}_t = \mathcal{F}_\infty^\Lambda \vee \mathcal{F}_t^X$ and the **F**-martingale

$$M(t) = \frac{e^{-r(u+X(t))}}{e^{\Lambda(t)h(r)-trc}}.$$

In Section 4.5.2 we considered the vector process (X, λ), the filtration **F** given by $\mathcal{F}_t = \mathcal{F}_t^\lambda \vee \mathcal{F}_t^X$, and an $\mathbf{F}^{(X,\,\lambda)}$-martingale of the form

$$M_u(t) = g(\lambda(t))e^{-R(u+X(t))},$$

where g is a positive function. The way to find g was to use Proposition 4.5 which says that if Y is a Markov process with generator A and v a function in the domain of A such that $Av \equiv 0$, then $v(Y(t))$ is an $\mathbf{F}^Y$-martingale.

Neither of these approaches seem quite applicable in the finite time case. In $M(t)$ the dependence of t is probably too complicated while in $M_u(t)$ we cannot vary r. We will here generalize the approach in Section 4.5.2.

We shall make use of the following special case* of an observation by Davis (1984, p. 370):

Let $Y = \{Y_t; t \geq 0\}$ be a (homogeneous) piecewise-deterministic Markov process with generator A, v a function in the domain of A, and ϑ a differentiable function such that

$$-\vartheta' \cdot v + Av \equiv 0.$$

Then M, defined by $M(t) = e^{-\vartheta(t)}v(Y_t)$, is an $\mathbf{F}^Y$-martingale.

INDICATION OF PROOF: We have

$$E^{\mathcal{F}_t^Y}[e^{-\vartheta(t+\Delta)}v(Y_{t+\Delta})] - e^{-\vartheta(t)}v(Y_t)$$
$$= e^{-\vartheta(t+\Delta)}\big(v(Y_t) + \Delta(Av)(Y_t) + o(\Delta)\big) - e^{-\vartheta(t)}v(Y_t)$$
$$= e^{-\vartheta(t)}\big[\big((1 - \Delta\vartheta(t) + o(\Delta))\big)\big(v(Y_t) + \Delta(Av)(Y_t) + o(\Delta)\big)\big)\big] - e^{-\vartheta(t)}v(Y_t)$$
$$= \Delta e^{-\vartheta(t)}v(Y_t)\big(-\vartheta'(t)v(Y_t) + \Delta(Av)(Y_t) + o(1)\big). \quad\blacksquare$$

Now we apply this to (X, λ), which is a piecewise-deterministic Markov process. For any fixed $r < r_\infty$ we look for a positive **F**-martingale ($\mathbf{F} = \mathbf{F}^{(X,\,\lambda)}$) of the form

$$M_u(t) = e^{-\theta(r)t}g(\lambda(t))e^{-r(u+X(t))},$$

where $g(0) = 1$ and $g(\ell) > 0$. It follows from Proposition 4.52 that M_u is an **F**-martingale if g and $\theta(r)$ satisfy

$$g(\ell)[\ell h(r) - rc - \theta(r)] + \eta(\ell)\int_0^\infty g(z)p_L(\ell, dz) - \eta(\ell)g(\ell) \equiv 0. \quad (22)$$

---

* At least when ϑ is monotone, this special case follows directly from Proposition 4.5 (Dynkin's theorem) applied to the vector process (ϑ, Y), which then is homogeneous. We will emphasize that the "Davis observation" applies to processes with more "genuine" inhomogeneity. Further Davis explicitly calculated the (generalized) generator.

When λ is an independent jump markovian intensity with state space $\mathbf{S}$, it follows – compare Theorem 4.53 – that M_u is an **F**-martingale for

$$g(\ell) = \frac{\eta(\ell)}{rc + \theta(r) + \eta(\ell) - h(r)\ell}$$

provided

(a) $\qquad rc + \theta(r) + \eta(\ell) - h(r)\ell > 0, \qquad p_L\text{-a.s.};$

(b) $\qquad \displaystyle\int_0^\infty \frac{\eta(\ell)}{rc + \theta(r) + \eta(\ell) - h(r)\ell} \, p_L(d\ell) = 1.$

If, for example, $\eta(\ell) \equiv 1$ and $p_L((\ell, \infty)) > 0$ for all $\ell > 0$ – compare Example 4.26 – there exists no $\theta(r)$ satisfying (a) for $r > 0$. Thus the *existence* of **F**-martingales of the required form is not guaranteed in this case. In order to avoid this and some other technical problems we *assume* that R – defined by Proposition 4.49 – is positive. Further we assume – for all $r \in [0, r_0)$, where r_0 is some value larger than R – that (a) and (b) have a differentiable and convex solution $\theta(r)$.

REMARK 10. We will indicate that the above assumptions are natural and ought to hold in "kind" cases. Since h is infinitely differentiable θ ought to be differentiable. The convexity seems natural since – see (4.86) – (a) and (b) are equivalent with

$$1 = \int_0^\infty \int_0^\infty e^{-s[rc+\theta(r)+\eta(\ell)-h(r)\ell]} \, \eta(\ell) \, ds \, p_L(d\ell).$$

Provided we may change the order of integration and differentiation we get

$$0 = \frac{d^2}{dr^2} \int_0^\infty \int_0^\infty e^{-s[rc+\theta(r)+\eta(\ell)-h(r)\ell]} \, \eta(\ell) \, ds \, p_L(d\ell)$$

$$= \frac{d}{dr} \int_0^\infty \int_0^\infty -s[c + \theta'(r) - h'(r)\ell] e^{-s[rc+\theta(r)+\eta(\ell)-h(r)\ell]} \, \eta(\ell) \, ds \, p_L(d\ell)$$

$$= \iint_0^\infty \left(s^2[c + \theta'(r) - h'(r)\ell]^2 - s[\theta''(r) - h''(r)\ell]\right) \cdot$$

$$\cdot e^{-s[rc+\theta(r)+\eta(\ell)-h(r)\ell]} \, \eta(\ell) \, ds \, p_L(d\ell),$$

which is possible only if $\theta''(r) > 0$.

When θ can be explitly found, we can – of course – directly check the assumptions. One such example is Case 2 in Section 4.6, where

$$p_L(d\ell) = \begin{cases} \frac{1}{2} & \text{if } 0 \leq \ell \leq 2 \\ 0 & \text{otherwise} \end{cases} \qquad \text{and} \qquad \eta(\ell) \equiv \eta.$$

Then we have

$$1 = -\frac{\eta}{2h(r)} \log\left(1 - \frac{2h(r)}{rc + \theta(r) + \eta}\right) \qquad \text{when } h(r) < \frac{rc + \theta(r) + \eta}{2}$$

and thus
$$\theta(r) = \frac{2h(r)}{1 - e^{-2h(r)/\eta}} - rc - \eta.$$

Obviously θ is differentiable. In order to check if θ is convex we consider functions a, b : $\mathbf{R}_+ \to \mathbf{R}$. If a and b are convex and $a' \geq 0$ then $a(b)$ is also convex, since
$$a(b)'' = a''(b)(b')^2 + a'(b)b'' \geq 0.$$

Put
$$a(x) = \frac{x}{1 - e^{-x}} \quad \text{and} \quad b(r) = \frac{2h(r)}{\eta}$$

and note that $\theta(r) = \eta \cdot a(b(r)) - rc - \eta$. Thus it is enough to check that a', $a'' \geq 0$. Straightforward derivation yields
$$a'(x) = \frac{1 - e^{-x} - xe^{-x}}{(1 - e^{-x})^2}, \qquad a'(0) = \frac{1}{2},$$

and
$$a''(x) = \frac{(x + 2 + xe^x - 2e^x)e^{-2x}}{(1 - e^{-x})^3}.$$

Since
$$x + 2 + xe^x - 2e^x = x + 2 + \sum_{k=0}^{\infty} \frac{x^{k+1}}{k!} - \sum_{k=0}^{\infty} \frac{2x^k}{k!}$$
$$= x + 2 + x + \sum_{k=2}^{\infty} \frac{kx^k}{k!} - 2 - 2x - \sum_{k=2}^{\infty} \frac{2x^k}{k!} \geq 0$$

the convexity follows.
□

We can now – in principle – proceed as in the classical and renewal cases. Nevertheless, since the martingale is slightly more complicated in this case, we give a detailed derivation.

Let $r < r_0$ be fixed and let $\theta(r)$ be the solution of (a) and (b). From (4.40) we get, in the "usual" way,
$$\Psi^{\mathcal{F}_0}(u, yu) \leq \frac{g(\lambda(0))e^{-ru}}{E^{\mathcal{F}_0}\left[e^{-\theta(r)T_u}g(\lambda(T_u)) \mid T_u \leq yu\right]}$$
$$\leq \frac{g(\lambda(0))e^{-ru}}{E^{\mathcal{F}_0}\left[\inf_{0 \leq t \leq yu} e^{-\theta(r)t}g(\lambda(t)) \mid T_u \leq yu\right]}$$

$$\leq \frac{g(\lambda(0))e^{-u\min(r,\,r-y\theta(r))}}{E^{\mathcal{F}_0}\left[\inf_{0\leq t\leq yu} g(\lambda(t)) \mid T_u \leq yu\right]}.$$

The problem – which was the reason for giving details – is that we, like in Theorem 4.53, must ensure that

$$\frac{1}{E^{\mathcal{F}_0}\left[\inf_{0\leq t\leq yu} g(\lambda(t)) \mid T_u \leq yu\right]} < \infty$$

and therefore we add the "additional" condition

(c) $\qquad \eta(\ell) \geq \beta, \qquad p_L\text{-a.s.} \qquad$ for some $\beta > 0$.

Then we have

$$\frac{1}{E^{\mathcal{F}_0}\left[\inf_{0\leq t\leq yu} g(\lambda(t)) \mid T_u \leq yu\right]} \leq \sup_{\ell \in \mathbf{S}} \frac{1}{g(\ell)}$$

$$= \sup_{\ell \in \mathbf{S}} \frac{rc + \theta(r) + \eta(\ell) - h(r)\ell}{\eta(\ell)} \leq 1 + \frac{rc + \theta(r)}{\beta}.$$

Since $\theta(R) = 0$, which follows from Proposition 4.30 (b), we get, exactly as in previous cases,

$$\Psi^{\mathcal{F}_0}(u, yu)$$

$$\leq \frac{\eta(\lambda(0))}{R_y c + \theta(R_y) + \eta(\lambda(0)) - h(R_y)\lambda(0)} \left(1 + \frac{R_y c + \theta(R_y)}{\beta}\right) e^{-R_y u},$$

where

$$R_y \overset{=}{_{>}} R \quad \text{as} \quad y \overset{\geq}{_{<}} \frac{1}{\theta'(R)} = y_0$$

and, for $y < y_0$,

$$R_y = f(r_y) \text{ where } r_y \text{ is the solution of } \theta'(r_y) = 1/y.$$

Exactly as in Theorem 4.53 we get the corresponding time-dependent Lundberg inequalities for $\Psi^0(u, yu)$ and $\Psi(u, yu)$.

Now we consider the case when λ is an M-state markovian intensity. Then (22) – compare Theorem 4.54 – is reduced to

$$[\mathbf{H}(r) - \theta(r)]\mathbf{g} = 0 \tag{23}$$

for some vector $\mathbf{g} > \mathbf{0}$. Any fixed $r > 0$, fulfilling (23), is the Lundberg exponent in a modified risk process with c replaced by $c + \frac{\theta(r)}{r}$. Since $r > 0$ we have, also in the modified risk process, positive safety loading. Thus Remark 4.46 is applicable, which means that 0 is an eigenvalue of $\mathbf{H}(r) - \theta(r)$ with maximal real part. Since, for any eigenvalue κ of $\mathbf{H}(r)$, $\kappa - \theta(r)$ is an eigenvalue of $\mathbf{H}(r) - \theta(r)$, it follows especially that $\theta(r) = \kappa(r)$. The "usual" martingale argument leads to $y_0 = 1/\kappa'(R)$ which is in agreement with (21).

A.4 Diffusion approximations

If little is known in the Cox case, nothing is known in the general case. The only method – known to us – which works for a very general class of underlying point processes is the "diffusion approximation." This approximation was discussed for the Poisson model in Section 1.2 and for the Cox model in Section 4.6. Recall, however, that its accuracy is not very good.

Let, as usual, the occurrence of the claims be described by a point process N. Assume that

$$\lim_{t \to \infty} \frac{\text{Var}[N(t)]}{t} = \sigma_N^2 \tag{24}$$

and that

$$N_n \xrightarrow{d} \sigma_N \cdot W \quad \text{as } n \to \infty, \tag{25}$$

where

$$N_n(t) = \frac{N(nt) - \alpha nt}{\sqrt{n}}$$

and W a standard Wiener process.

Strictly speaking, only (25) is necessary for the diffussion approximation, but with (24) σ_N gets a natural interpretation. The assumptions do not seem too restrictive. We have seen that they hold in the Poisson case with $\sigma_N^2 = \alpha$ and in the Cox case, compare Section 4.6, with $\sigma_N^2 = \alpha + \sigma_\Lambda^2$. In the renewal case, see Billingsley (1968, p. 148), they hold with $\sigma_N^2 = \sigma_S^2 \alpha^3$, where σ_S^2 is the variance of the inter-occurrence time distribution.

Define $\bar{S}_n$ by

$$\bar{S}_n(t) = \frac{\bar{S}(nt) - \alpha \mu nt}{\sqrt{n}},$$

where $\bar{S}(t) = \sum_{k=1}^{N(t)} Z_k$. Then, see for example Grandell (1977, p. 47),

$$\bar{S}_n \xrightarrow{d} \sqrt{\mu^2 \sigma_N^2 + \alpha \sigma^2} \cdot W \quad \text{as } n \to \infty. \tag{26}$$

REMARK 11. In order to make (26) probable, we will indicate the proof of its "one-dimensional version," i.e., that

$$\bar{S}_n(t) \xrightarrow{d} \sqrt{\mu^2 \sigma_N^2 + \alpha \sigma^2} \cdot W(t) \quad \text{as } n \to \infty.$$

Put $\tilde{S}(k) = \sum_{j=1}^{k} Z_j$ and note that $\bar{S}(t) = \tilde{S}(N(t))$. We have

$$\bar{S}_n(t) = \frac{\tilde{S}(N(nt)) - \alpha \mu nt}{\sqrt{n}}$$

$$= \sigma \cdot \sqrt{\frac{N(nt)}{n}} \cdot \frac{\tilde{S}(N(nt)) - \mu N(nt)}{\sigma \sqrt{N(nt)}} + \mu \cdot \frac{N(nt) - \alpha nt}{\sqrt{n}}$$

$$\xrightarrow{d} \sigma \sqrt{\alpha t} \cdot W_1(1) + \mu \sigma_N \cdot W_2(t) \stackrel{d}{=} \sigma \sqrt{\alpha} \cdot W_1(t) + \mu \sigma_N \cdot W_2(t)$$

$$\stackrel{d}{=} \sqrt{\alpha\sigma^2 + \mu^2\sigma_N^2} \cdot W(t),$$

where W_1 and W_2 are independent standard Wiener processes. The notation $\stackrel{d}{=}$ means "equality in distribution."

□

Define Y_n and Y by

$$Y_n(t) = \frac{c_n nt - \bar{S}(nt)}{\sqrt{n}} \quad \text{and} \quad Y(t) = \gamma\alpha\mu t - \sqrt{\mu^2\sigma_N^2 + \alpha\sigma^2} \cdot W(t),$$

Then, see Section 4.6,

$$Y_n \stackrel{d}{\to} Y \quad \text{as} \quad n \to \infty$$

if and only if

$$\rho_n\sqrt{n} = \frac{c_n - \alpha\mu}{\alpha\mu}\sqrt{n} \to \gamma \quad \text{as} \quad n \to \infty.$$

Recall from the survey "Basic facts about weak convergence" in Section 1.2 that $Y_n \stackrel{d}{\to} Y$ implies

$$\inf_{0 \leq t \leq t_0} Y_n(t) \stackrel{d}{\to} \inf_{0 \leq t \leq t_0} Y(t) \quad \text{for any } t_0 < \infty$$

but not necessarily $\inf_{t \geq 0} Y_n(t) \stackrel{d}{\to} \inf_{t \geq 0} Y(t)$.

Define $\Psi_n^D(u_0, t_0)$ and $\Psi^D(u_0, t_0)$ by

$$\Psi_n^D(u_0, t_0) = P\{\inf_{0 \leq t \leq t_0} Y_n(t) < -u_0\}$$

and

$$\Psi^D(u_0, t_0) = P\{\inf_{0 \leq t \leq t_0} Y(t) < -u_0\}.$$

Then

$$\Psi_n^D(u_0, t_0) \to \Psi^D(u_0, t_0) \quad \text{as} \quad n \to \infty$$

and, see (9),

$$\Psi^D(u_0, t_0)$$
$$= 1 - \mathcal{N}\left(\frac{u_0 + \gamma\alpha\mu t_0}{\sqrt{(\mu^2\sigma_N^2 + \alpha\sigma^2)t_0}}\right) + \mathcal{N}\left(\frac{-u_0 + \gamma\alpha\mu t_0}{\sqrt{(\mu^2\sigma_N^2 + \alpha\sigma^2)t_0}}\right) e^{-Ru_0}, \quad (27)$$

where

$$R = \frac{2\gamma\alpha\mu}{\mu^2\sigma_N^2 + \alpha\sigma^2}.$$

Consider now a risk process X with relative safety loading ρ and the corresponding ruin probability $\Psi(u, t)$. Then we have, for each n,

$$\Psi(u, t) = P\{\inf_{0 \leq s \leq t} X(s) < -u\} = P\{\inf_{0 \leq s \leq t} cs - \bar{S}(s) < -u\}$$

A.4 Diffusion approximations

$$= P\left\{\inf_{0\leq s\leq t} \frac{cs - \bar{S}(s)}{\sqrt{n}} < -\frac{u}{\sqrt{n}}\right\} = P\left\{\inf_{0\leq s\leq t/n} \frac{cns - \bar{S}(ns)}{\sqrt{n}} < -\frac{u}{\sqrt{n}}\right\}.$$

Assume now that ρ is small, u is large, and t is very large in such a way that ρ^{-1}, u, and $\sqrt{t}$ are of the same large order. Put

$$\gamma = \rho\sqrt{n}, \quad u_0 = \frac{u}{\sqrt{n}}, \quad \text{and} \quad t_0 = \frac{t}{n} \tag{28}$$

where n is chosen such that γ, u_0, and t_0 are of the same moderate order. This leads to the diffusion approximation

$$\Psi(u,t) \approx \Psi^D(u_0, t_0) = \Psi_D(u,t), \tag{29}$$

where

$$\Psi_D(u,t) = 1 - \mathcal{N}\left(\frac{u + \rho\alpha\mu t}{\sqrt{(\mu^2\sigma_N^2 + \alpha\sigma^2)t}}\right) + \mathcal{N}\left(\frac{-u + \rho\alpha\mu t}{\sqrt{(\mu^2\sigma_N^2 + \alpha\sigma^2)t}}\right)e^{-R_D u}$$

and

$$R_D = \frac{2\rho\alpha\mu}{\mu^2\sigma_N^2 + \alpha\sigma^2}.$$

As an illustration of (29) we consider $\rho = 5\%$, $u = 100$, and $t = 1000$ in the four combinations of exponentially/mixed exponentially distributed claims and Poisson/renewal case discussed in Examples 1, 4, and 6. In all cases we have $\alpha = \mu = 1$. Note that the diffusion approximation does not differ in the ordinary and stationary case. Then we have

$\sigma^2 = 1$ for exponentially distributed claims;
$\sigma^2 = 42.1982$ for mixed exponentially distributed claims;*
$\sigma_N^2 = 1$ in the Poisson case;
$\sigma_N^2 = 2.5$ in the renewal case.**

TABLE 7. Illustration of the diffusion approximation. The values of σ^2 and σ_N^2 indicate the model.

σ^2	σ_N^2	$\Psi(100, 1000)$	$\Psi_D(100, 1000)$
1	1	0.0019	0.0013
42.1982	1	0.4115	0.5565
1	2.5	0.0196	0.0170
42.1982	2.5	0.4209	0.5640

* $F(z) = 1 - 0.0039793e^{-0.014631z} - 0.1078392e^{-0.190206z} - 0.8881815e^{-5.514588z}$.
** $K^0(t) = 1 - 0.25e^{-0.4t} - 0.75e^{-2t}$.

From Table 7 it is seen that the accuracy of the diffusion approximation is, as was to be expected, not very good.

Define, for any $z \in D$ and any $u \geq 0$ the function $t_u : D \to [0, \infty]$ by

$$t_u(z) = \inf\{s \geq 0 \mid z(s) < -u\}.$$

Note that $t_u(X) = T_u$. Further,

$$Y_n \xrightarrow{d} Y \quad \text{implies} \quad t_u(Y_n) \xrightarrow{d} t_u(Y). \tag{30}$$

For γ, u_0, and t_0 given by (28) and X_n defined by $X_n(t) = X(nt)/\sqrt{n}$ we get

$$t_{u_0}(X_n) = \inf\{s \geq 0 \mid X(ns) < u_0\sqrt{n}\} = \frac{1}{n}t_{u_0\sqrt{n}}(X) = \frac{1}{n}t_u(X).$$

Thus we have

$$\frac{1}{u}t_u(X) = \frac{\sqrt{n}}{u_0}t_{u_0}(X_n) \quad \text{and} \quad \frac{1}{\sqrt{u}}t_u(X) = \sqrt{\frac{n^{3/2}}{u_0}}t_{u_0}(X_n). \tag{31}$$

Put, see (9) and the definition of Y,

$$y_0^D \stackrel{\text{def}}{=} \frac{1}{\alpha\gamma\mu} = \frac{1}{\alpha\mu\rho\sqrt{n}} \quad \text{and} \quad v_0^D \stackrel{\text{def}}{=} \frac{\mu^2\sigma_N^2 + \alpha\sigma^2}{(\alpha\gamma\mu)^3} = \frac{\mu^2\sigma_N^2 + \alpha\sigma^2}{(\alpha\mu\rho)^3 n^{3/2}}.$$

Equations (29), (30), and (31) lead to the approximations

$$y_0 \approx y_{0_D} \stackrel{\text{def}}{=} \frac{1}{\alpha\mu\rho} \quad \text{and} \quad v_0 \approx v_{0_D} \stackrel{\text{def}}{=} \frac{\mu^2\sigma_N^2 + \alpha\sigma^2}{(\alpha\mu\rho)^3}.$$

As an illustration we consider the same cases as in Table 7.

TABLE 8. Illustration of the diffusion approximation for $\rho = 5\%$. The values of σ^2 and σ_N^2 indicate the model.

σ^2	σ_N^2	y_0	y_{0_D}	$\sqrt{v_{0_D}}$	$\sqrt{v_0}$
1	1	19.05	20	126.49	126.49
42.1982	1	17.51	20	587.84	587.87
1	2.5	19.35	20	167.33	167.33
42.1982	2.5	17.60	20	597.96	597.98

The values of y_0 in Table 8 indicate that y_{0_D} works reasonably well for $\rho = 5\%$. This is also true in the Cox case when $\lambda(t)$ is the two-state Markov process illustrated in Table 6. The accuracy of v_{0_D} is almost perfect. Generally speaking, the approximations y_{0_D} and v_{0_D} seem to work better than

A.4 Diffusion approximations

expected from the poor accuracy of the diffusion approximations of ruin probabilities.

In Section 4.6 we had a similar experience when using the diffusion approximation as a motivation for certain approximations of the Lundberg exponent. In that case those approximations also worked rather well for larger values of ρ. It is tempting to see if this is also the case here, and therefore we consider in Table 9 $\rho = 20\%$.

TABLE 9. Illustration of the diffusion approximation for $\rho = 20\%$. The values of σ^2 and σ_N^2 indicate the model.

σ^2	σ_N^2	y_0	y_{0_D}	$\sqrt{v_{0_D}}$	$\sqrt{v_0}$
1	1	4.17	5	15.81	15.81
42.1982	1	3.10	5	73.25	73.48
1	2.5	4.44	5	20.92	20.92
42.1982	2.5	3.15	5	74.51	74.75

Table 9 indicates that y_{0_D} does not work so well for $\rho = 20\%$ but the accuracy of v_{0_D} is still almost perfect. Out of sheer curiosity – this is probably only of little interest – we consider in Table 10 $\rho = 100\%$. The figures do not require any comments.

TABLE 10. Illustration of the diffusion approximation for $\rho = 100\%$. The values of σ^2 and σ_N^2 indicate the model.

σ^2	σ_N^2	y_0	y_{0_D}	$\sqrt{v_{0_D}}$	$\sqrt{v_0}$
1	1	0.50	1	1.41	1.41
42.1982	1	0.23	1	6.37	6.57
1	2.5	0.65	1	1.90	1.87
42.1982	2.5	0.23	1	6.47	6.69

Let us go back and consider the Poisson case. It follows from (14) that $y_0 \leq y_{0_D}$. This indicates that the our final remarks in that case, which were based on (14), seem to hold more generally.

REMARK 12. In the Poisson case we exploited the De Vylder approximation. Applying that idea to y_0 – rather than to $\alpha \cdot uy_0$ – and v_0 we get

$$\tilde{y}_0 = \frac{1}{\tilde{\alpha}\tilde{\mu}\tilde{\rho}(1+\tilde{\rho})} = \frac{1}{\alpha\mu\rho(1+\frac{2\mu\zeta_3}{3\zeta_2^2}\rho)} \quad \text{and} \quad \tilde{v}_0 = \frac{2}{\tilde{\alpha}^2\tilde{\mu}\tilde{\rho}^3} = \frac{\zeta_2}{\alpha^2\mu^3\rho^3} = v_{0_D}.$$

The fact that $\tilde{v}_0 = v_{0_D}$ may partially explain why v_{0_D} seems to also work so well for larger values of ρ.
□

REMARK 13. Some readers are perhaps puzzled by the fact that $Y_n \xrightarrow{d} Y$ implies $t_u(Y_n) \xrightarrow{d} t_u(Y)$ but not $\inf_{t \geq 0} Y_n(t) \xrightarrow{d} \inf_{t \geq 0} Y(t)$. We will therefore consider weak convergence in slightly more detail than we did in Section 1.2.

Recall that D is the space of right-continuous functions with left-hand limits endowed with the Skorohod J_1 topology. Let C denote the subspace of continuous functions. Let $x, x_1, x_2, \ldots$ be functions in D. For $x \in C$ the convergence $x_n \to x$ is equivalent to

$$\lim_{n \to \infty} \sup_{0 \leq s \leq t_0} |x_n(s) - x(s)| = 0 \quad \text{for all } t_0 < \infty.$$

If $x \notin C$ the definition of $x_n \to x$ is more complicated in that sense that a sequence of time transformations is introduced.

Define, for any $z \in D$ and any $t_0 < \infty$, the functions i_{t_0} and $i : D \to [-\infty, \infty]$ by

$$i_{t_0}(z) = \inf_{0 \leq s \leq t_0} z(s) \quad \text{and} \quad i(z) = \inf_{s \geq 0} z(s).$$

The function i_{t_0} is continuous on C (and on D):
Put $d_n = \sup_{0 \leq s \leq t_0} |x_n(s) - x(s)|$. Since $x_n(s) \geq x(s) - d_n \geq i_{t_0}(x)$, for $0 \leq s \leq t_0$, we have $i_{t_0}(x_n) \geq i_{t_0}(x) - d_n$ or $i_{t_0}(x) - i_{t_0}(x_n) \leq d_n$. Similarly $i_{t_0}(x_n) - i_{t_0}(x) \leq d_n$ and thus $|i_{t_0}(x_n) - i_{t_0}(x)| \leq d_n$.

The function i is not continuous on C:
Consider

$$x_n(t) = \begin{cases} 0 & \text{if } t < n \\ -1 & \text{if } t \geq n \end{cases} \quad \text{and} \quad x(t) \equiv 0.$$

Then $i(x_n) = -1$ and $i(x) = 0$ while $x_n \to x$.

The function t_u, for $u \geq 0$, is not continuous on C:
Consider

$$x_n(t) = \begin{cases} -t & \text{if } t < u + \frac{1}{n} \\ -u - \frac{1}{n} & \text{if } t \geq u + \frac{1}{n} \end{cases} \quad \text{and} \quad x(t) = \begin{cases} -t & \text{if } t < u \\ -u & \text{if } t \geq u \end{cases}.$$

Then $t_u(x_n) = u$ and $t_u(x) = \infty$ while $x_n \to x$.

Let X_1, $X_2, \ldots$ be processes in D with $X_n(0) = 0$ and X a Wiener process with positive drift: $X(t) = \beta t + \delta W(t)$, $\beta > 0$. Denote the distribution of X_n (X) by P_n (P) and note that $P\{C\} = 1$.

The "main theorem of weak convergence" states that $X \xrightarrow{d} X$ implies $f(X_n) \xrightarrow{d} f(X)$ for any measurable and P-a.s. continuous function f. Especially this means that it is enough to show that f is continuous on C when the limit process X is in C.

Assume that $X_n \xrightarrow{d} X$. Then $i_{t_0}(X_n) \xrightarrow{d} i_{t_0}(X)$ which, in turn, implies $P_n\{i_{t_0}(X_n) < u\} \to P\{i_{t_0}(X) < u\}$ for those u where $P\{i_{t_0}(X) = u\} = 0$. For a Wiener process with positive drift this, see (9), holds in fact for all $u < \infty$. Since

$$\{z \in D \mid t_u(z) \leq t_0\} = \{z \in D \mid i_{t_0}(z) < -u\}$$

it follows that $P_n\{t_u(X_n) \leq t_0\} \to P\{t_u(X) \leq t_0\}$. Since this argument goes through for all $t_0 < \infty$ (30) follows. Equation (30) does, however, not imply that $P_n\{t_u(X_n) < \infty\} \to P\{t_u(X) < \infty\}$ since $P\{t_u(X) = \infty\} > 0$.

Note that (30) followed from properties of the Wiener process. These properties are not enough to guarantee that $i(X_n) \xrightarrow{d} i(X)$, as seen by the following example:
Put

$$X_n(t) = \begin{cases} X(t) & \text{if } t < n \\ -t & \text{if } t \geq n \end{cases}$$

which implies that $X_n \xrightarrow{d} X$. This statement really does require some more properties of weak convergence in D than given here. In fact, a basic result in Lindvall (1973) is that convergence in distribution of processes on $[0, \infty)$ can be brought back to processes restricted to $[0, t_k]$, $k = 1, 2, \ldots$, such that $t_k \to \infty$. The reader may believe this from our discussion of convergence in D or – better – consult Lindvall (1973).

Obviously $i(X_n) = -\infty$ while $P\{i(X) > -\infty\} = 1$. Further

$$t_u(X_n) = \min[t_u(X), \max(u, n)] \to t_u(X)$$

while

$$1 = P\{t_u(X_n) < \infty\} \not\to P\{t_u(X) < \infty\} < 1.$$

If we are to prove $i(X_n) \xrightarrow{d} i(X)$ we therefore must use some special properties of X_n. As mentioned in Section 1.2, this can be done for Y_n, as defined above, in the Poisson case. Our conjecture is that $i(Y_n) \xrightarrow{d} i(Y)$ holds rather generally. An argument for this is that Y_n contains a contraction of time, while our counterexamples are based on a drift of the time of ruin to infinity.

Finally we will emphasize that, although the approximations y_{0_D} and v_{0_D} were *motivated* by weak convergence, they do not *follow* from any limit theorem. The "parameters" y_0 and v_0 are defined by limit theorems as (3) and (4) where the limit procedure – $u, t \to \infty$ and $t = O(u)$ – is different from the one here. Such limit theorems are, furthermore, only known for special models. Therefore y_{0_D} and v_{0_D} must be looked upon as based on ad hoc reasoning.

References and author index

ALLEN, A. (1978) *Probability, Statistics and Queueing Theory with Computer Science Applications.* Academic Press, New York. [126, 127]*

ALMER, B. (1957) Risk analysis in theory and practical statistics. *Transactions XVth International Congress of Actuaries, New York,* II, 314 - 370. [55]

AMMETER, H. (1948) A generalization of the collective theory of risk in regard to fluctuating basic probabilities. *Skand. AktuarTidskr.*, 171 - 198. [77, 104, 105, 119]

ANDERSEN, E. SPARRE (1957) On the collective theory of risk in the case of contagion between the claims. *Transactions XVth International Congress of Actuaries, New York,* II, 219 - 229. [57, 60, 61, 75]

ARFWEDSON, G. (1955) Research in collective risk theory. Part 2. *Skand. AktuarTidskr.*, 53 - 100. Part 1 in *SAT* (1954, pp. 191 - 223.) [137, 138]

ASMUSSEN, S. (1984) Approximations for the probability of ruin within finite time. *Scand. Actuarial J.*, 31 - 57. Erratum in *SAJ* (1985, p. 64). [25, 139]

ASMUSSEN, S. (1985) Conjugate processes and the simulation of ruin problems. *Stochastic Proc. Applic.* 20, 213 - 229. [15]

ASMUSSEN, S. (1987) *Applied Probability and Queues.* John Wiley & Sons, New York. [123]

ASMUSSEN, S. (1989) Risk theory in a markovian environment. *Scand. Actuarial J.*, 66 - 100. [77, 117, 119, 152, 153]

VON BAHR, B. (1974) Ruin probabilities expressed in terms of ladder height distributions. *Scand. Actuarial J.*, 190 - 204. [145]

BEEKMAN, J. (1969) A ruin function approximation. *Trans. of the Soc. of Actuaries 21,* 41 - 48 and 275 - 279. [18]

BENCKERT, L.-G. and JUNG, J. (1974) Statistical models of claim distribution in fire insurance. *Astin Bulletin VII,* 1 - 25. [23]

---

* Pages on which references are cited are given in brackets.

BERG, C. (1981) The Pareto distribution is a generalized Γ-convolution – a new proof. *Scand. Actuarial J.*, 117 - 119.   [48]

BILLINGSLEY, P. (1968) *Convergence of Probability Measures.* John Wiley & Sons, New York.   [15, 159]

BJÖRK, T. and GRANDELL, J. (1985) An insensitivity property of the ruin probability. *Scand. Actuarial J.*, 148 - 156.   [125, 127]

BJÖRK, T. and GRANDELL, J. (1988) Exponential inequalities for ruin probabilities in the Cox case. *Scand. Actuarial J.*, 77 - 111.
[77, 92, 95, 99, 100, 102 – 105, 108, 109, 112, 114, 116]

BRÉMAUD, P. (1972) *A Martingale Approach to Point Processes.* Ph.D. Thesis, Memo ERL-M345, Dept. of EECS, Univ. of Calif., Berkeley.
[38, 40]

BRÉMAUD, P. (1981) *Point Processes and Queues. Martingale Dynamics.* Springer-Verlag, New York.   [38]

CRAMÉR, H. (1930) *On the Mathematical Theory of Risk.* Skandia Jubilee Volume, Stockholm.   [4, 13]

CRAMÉR, H. (1945) *Mathematical Methods of Statistics.* Almqvist & Wiksell, Stockholm and Princeton University Press, Princeton.   [19, 27]

CRAMÉR, H. (1955) *Collective Risk Theory.* Skandia Jubilee Volume, Stockholm.   [vi, vii, 4, 13, 21, 35, 65, 67, 140]

DALEY, D. J. and VERE-JONES, D. (1988) *An Introduction to the Theory of Point Processes.* Springer-Verlag, New York.   [41]

DASSIOS, A. and EMBRECHTS, P. (1989) Martingales and insurance risk. *Commun. Statist. – Stochastic models* 5, 181 - 217.   [vi, 146, 147]

DAVIS, M. H. A. (1984) Piecewise-deterministic Markov processes: A general class of non-diffusion stochastic models. *J. R. Statist. Soc. B 46*, 353 - 388.   [147, 155]

DELBAEN, F. and HAEZENDONCK, J. (1987) Classical risk theory in an economic environment. *Insurance: Mathematics and Economics* 6, 85 - 116.   [vi]

DE VYLDER, F. (1977) Martingales and ruin in a dynamical risk process. *Scand. Actuarial J.*, 217 - 225.   [40]

DE VYLDER, F. (1978) A practical solution to the problem of ultimate ruin probability. *Scand. Actuarial J.*, 114 - 119.   [19, 20, 24]

ELLIOTT, R. J. (1982) *Stochastic Calculus and Applications.* Springer-Verlag, New York.   [9, 39]

EMBRECHTS, P. and VERAVERBEKE, N. (1982) Estimates for the probability of ruin with special emphasis on the possibility of large claims. *Insurance: Mathematics and Economics* 1, 55 - 72.   [23]

FELLER, W. (1971) *An Introduction to Probability Theory and its Applications. Vol II.* 2nd ed. John Wiley & Sons, New York.
[4, 6, 30, 52, 62 – 64, 66, 77, 79, 81, 125]

FRANKEN, P., KÖNIG, D., ARNDT, U., and SCHMIDT, V. (1981) *Queues and Point Processes.* Akademie-Verlag, Berlin and John Wiley & Sons, New York. [41, 96, 107, 110, 127, 128, 130]

GERBER, H. U. (1973) Martingales in risk theory. *Mitt. Ver. Schweiz. Vers. Math. 73*, 205 - 216. [8, 136]

GERBER, H. U. (1979) *An Introduction to Mathematical Risk Theory.* S. S. Heubner Foundation monograph series 8, Philadelphia. [vi, 14, 37]

GRANDELL, J. (1976) *Doubly Stochastic Poisson Processes.* Lecture Notes in Math. 529, Springer-Verlag, Berlin [35, 36, 43, 47, 122]

GRANDELL, J. (1977) A class of approximations of ruin probabilities. *Scand. Actuarial J. Suppl.*, 38 - 52. [16, 20, 25, 122, 159]

GRANDELL, J. (1978) A remark on 'A class of approximations of ruin probabilities.' *Scand. Actuarial J.*, 77 - 78. [16, 20]

GRANDELL, J. (1979) Empirical bounds for ruin probabilities. *Stochastic Proc. Applic. 8*, 243 - 255. [25]

GRANDELL, J. and SEGERDAHL, C.-O. (1971) A comparison of some approximations of ruin probabilities. *Skand. AktuarTidskr.*, 144 - 158. [14, 20, 21, 74]

GRIGELIONIS, B. (1963) О сходимости сумм ступенчатых случайных процессов к пуассоновскому. «Теор. вероят. и примен.» 8, 189 - 194. English translation: On the convergence of step processes to a Poisson process. *Theor. Prob. Appl. 8*, 177 - 182. [44]

GRIGELIONIS, B. (1975) Случайные точечные процессы и мартингалы. *Liet. Matem. Rink. 15*, 101 - 114. English translation: Random point processes and martingales. *Lithuanian Math. J. 15*, 444 - 453. [40]

HABERLAND, E. (1976) Infinitely divisible stationary recurrent point processes. *Math. Nachr. 70*, 259 - 264. [45]

HERKENRATH, U. (1986) On the estimation of the adjustment coefficient in risk theory by means of stochastic approximation procedures. *Insurance: Mathematics and Economics 5*, 305 - 313. [31, 32]

HÖGLUND, T. (1990) An asymptotic expression for the probability of ruin within finite time. *Ann. Prob. 18*, 378 - 389. [145]

IGLEHART, D. L. (1969) Diffusion approximations in collective risk theory. *J. Appl. Prob. 6*, 285 - 292. [16]

KALLENBERG, O. (1975) Limits of compound and thinned point processes. *J. Appl. Prob. 12*, 269 - 278. [46]

KALLENBERG, O. (1983) *Random Measures*. 3rd ed. Akademie-Verlag, Berlin and Academic Press, New York. [41, 44]

KARR, A. F. (1986) *Point Processes and their Statistical Inference*. Marcel Dekker, New York. [41, 107]

KINGMAN, J. F. C. (1964) On doubly stochastic Poisson processes. *Proc. Camb. Phil. Soc. 60*, 923 - 930. [47, 70]

KINGMAN, J. F. C. (1972) *Regenerative Phenomena*. John Wiley & Sons, New York. [47]

KOLSRUD, T. (1986) Some comments on thinned renewal processes. *Scand. Actuarial J.*, 136 - 241. [53]

KUMMER, G. and MATTHES, K. (1970) Verallgemeinerung eines Satzes von Sliwnjak III. *Rev. Roum. Math. Pure et Appl. 15*, 1631 - 1642. [131]

LINDVALL, T. (1973) Weak convergence of probability measures and random functions in the function space $D[0,\infty)$. *J. Appl. Prob. 10*, 109 - 121. [15, 165]

LIPTSER, R. S. and SHIRYAYEV, A. N. (1978) *Statistics of Random Processes II. Applications*. Springer-Verlag, New York. [38, 39]

LUNDBERG, F. (1903) *I. Approximerad Framställning av Sannolikhetsfunktionen. II. Återförsäkring av Kollektivrisker*. Almqvist & Wiksell, Uppsala. [13]

LUNDBERG, F. (1926) *Försäkringsteknisk Riskutjämning*. F. Englunds boktryckeri A.B., Stockholm. [13]

MATTHES, K., KERSTAN, J., and MECKE, J. (1978) *Infinitely Divisible Point Processes*. John Wiley & Sons, New York. [41, 44, 129, 131]

MECKE, J. (1968) Eine charakteristische Eigenschaft der doppelt stochastischen Poissonschen Prozesse. *Z. Wahrschein. und Verw. Geb. 11*, 74 - 81. [46]

REINHARD, J. M. (1984) On a class of semi-Markov risk models obtained as classical risk models in a markovian environment. *Astin Bulletin XIV*, 23 - 43. [77, 84, 86 – 88]

RÉNYI, A. (1960) On the central limit theorem for the sum of a random number of independent random variables. *Acta Math. Acad. Sci. Hung. 11*, 97 - 102. [28]

REUTER, G. E. H. (1956) Über eine Volterrasche Integralgleichung mit totalmonotonem Kern. *Arch. Math. 7*, 59 - 66. [49]

ROSENLUND, S. (1989) Numerical calculation of the Cramér-Lundberg approximation. *Scand. Actuarial J.*, 119 - 122. [30]

SACKS, J. (1958) Asymptotic distribution of stochastic approximation procedures. *Ann. Math. Stat. 22*, 373 - 405. [32]

SEAL, H. L. (1969) Simulation of the ruin potential of nonlife insurance companies. *Trans. of the Soc. of Actuaries 21*, 563 - 585. [14]

SEGERDAHL, C.-O. (1955) When does ruin occur in the collective theory of risk. *Skand. AktuarTidskr.*, 22 - 36. [137, 138]

SERFOZO, R. (1972) Processes with conditionally independent increments. *J. Appl. Prob. 9*, 303 - 315. [40]

SVENSSON, Å. (1987) Some Inequalities for Thinned Point Processes. Dept. of Actuar. Math. and Mathematical Statistics, Univ. of Stockholm, Research report. [51]

TAKÁCS, L. (1962) *Introduction to the Theory of Queues.* Oxford University Press, New York. [127]

THEDÉEN, T. (1986) The Inverses of Thinned Point Processes. Dept. of Statistics, Univ. of Stockholm, Research report (1986:1). [46]

THORIN, O. (1971) Further remarks on the ruin problem in case the epochs of the claims form a renewal process. *Skand. AktuarTidskr.*, 14 - 38 and 121 - 142. [152]

THORIN, O. (1973) The ruin problem in case the tail of the claim distribution is completely monotone. *Skand. AktuarTidskr.*, 100 - 119. [14, 52]

THORIN, O. (1974) On the asymptotic behavior of the ruin probability for an infinite period when the epochs of claims form a renewal process. *Scand. Actuarial J.*, 81 - 99. [57, 65]

THORIN, O. (1975) Stationarity aspects of the Sparre Andersen risk process and the corresponding ruin probabilities. *Scand. Actuarial J.*, 87 - 98. [69, 70]

THORIN, O. (1977) Ruin probabilities prepared for numerical calculations. *Scand. Actuarial J. Suppl.*, 7 - 17. [14]

THORIN, O. (1982) Probabilities of ruin. *Scand. Actuarial J.*, 65 - 102. [57]

THORIN, O. (1986) Ruin probabilities when the claim amounts are gamma distributed. Försäkringstekniska forskningsnämnden. Meddelande nr 69. [14]

THORIN, O. (1988) Personal communications. [48]

THORIN, O. and WIKSTAD, N. (1973) Numerical evaluation of ruin probabilities for a finite period. *Astin Bulletin VII*, 137 - 153. [75, 76]

THORIN, O. and WIKSTAD, N. (1977) Calculation of ruin probabilities when the claim distribution is lognormal. *Astin Bulletin IX*, 231 - 246. [14, 23]

WATANABE, S. (1964) On dicontinuous additive functionals and Lévy measures of a Markov process. *Japan. J. Math. 34*, 53 - 70. [38]

WIKSTAD, N. (1971) Examplification of ruin probabilities. *Astin Bulletin VI*, 147 - 152.   [21, 75, 139, 143, 150, 151]

WIKSTAD, N. (1983) A numerical illustration of differences between ruin probabilities originated in the ordinary and in the stationary cases. *Scand. Actuarial J.*, 47 - 48.   [75 76]

YANNAROS, N. (1985)   On the Thinning of Renewal Point Processes. Dept. of Statistics, Univ. of Stockholm, Research report (1985:6).   [53]

YANNAROS, N. (1988a)   On Cox processes and gamma renewal processes. *J. Appl. Prob.* 25, 423 - 427.   [52]

YANNAROS, N. (1988b)   The inverses of thinned renewal processes. *J. Appl. Prob.* 25, 822 - 828.   [50, 51]

Subject index

A
Adapted to **F**   9
approximation
  Beekman-Bowers –   18
  Cramér-Lundberg –   7, 65, 69, 88, 151
  De Vylder –   19, 142, 163
  Diffusion –   17, 122, 162
Ascending ladder point   62
Associated random walk   65

B
Beekman-Bowers approximation   18
Borel algebra   41
Borel measure   41

C
Chapman-Kolmogorov equations   78
Classical risk process   4
Coefficient of variation   55
Compensator   39
Completely monotone   52
Convergence in distribution   15
Cox process   35, 39
Cramér-Lundberg approximation
  Poisson case   7
  renewal case   65, 69
  special Cox cases   88, 151

D
De Vylder approximation   19, 142, 163
Descending ladder point   63
Differential argument   4
Diffuse random measure   35
Diffusion approximation   17, 122, 161
Distribution of a random measure   42
Doubly stochastic Poisson process   35
Dynkin's theorem   79

E
Ergodic point process   129
Exponentially distributed claims
  Poisson case   5, 6, 12, 17, 18, 30, 138
  renewal case   60, 66, 69
  special Cox cases   85 – 92, 118, 120 – 124, 152

F
F-compensator   39
F-Cox process   39
F-martingale   9
F-Poisson process   38
F-stopping time   9
F-supermartingale   9
Filtration   9

G
Generator 79
Gross risk premium rate 1

H
Homogeneous Markov process 78

I
Independent jump intensity 95
Individual intensity 128
Infinitely divisible
  point process 44
  random measure 44
Infinitesimal operator 79
Initial distribution 78
Intensity for
  an exponential distribution 48
  a random measure 43
Intensity function for
  a markovian jump process 80
  a Poisson process 34
Intensity matrix 83
Intensity measure for a random measure 42
Intensity process for a Cox process 36
Invariant sets 129

J
Jump measure for a markovian jump process 80

L
Laplace transform 13, 52
Life annuity insurance 8, 12, 149
Lundberg exponent
  Cox case 94
  Poisson case 7
  renewal case 59
  special Cox cases 88, 103, 109, 116, 117
  time-dependent 136, 148, 158
Lundberg inequality
  Cox case 94
  Poisson case 11
  renewal case 60, 69
  special Cox cases 88, 103, 104, 109, 116, 117
  time-dependent 136, 148, 158

M
M-state Markov process 82
Martingale 9
Martingale approach 10
Markov process 78
  Homogeneous – 78
  Two-state – 48
  with independent jumps 82
Markov renewal intensity 106
Markovian jump process 80
Maximal eigenvalue 108
Mixed Poisson process 43

N
Negative risk sums 8
Non-arithmetic distribution 61

O
Operational time scale 35
Ordinary
  independent jump intensity 96
  renewal process 43, 57

P
p-inverse 45
p-thinning 45
p-thinning operator 45
Palm distribution 130
Palm process 130
Phase space 41
Point process 41
  Simple – 34
  with intensity α 1

Subject index 175

Poisson model 4
Poisson process
  Standard – 34
  with intensity α 4
  with intensity measure A 33
Polish space 15
Positive risk sums 8
Positive safety loading 2, 58

R
Random measure 35, 42
Random walk 59
Reduced Palm process 131
Relative safety loading 1, 58, 98
Renewal argument 4
Renewal measure 62
Renewal process 43, 57
Right continuous **F**-(super)martingale 9
Risk process 1
Ruin probability 1
  Finite-time – 136

S
Safety loading 1
Shift operator 43
σ-algebra strictly prior to ruin 12
Simple point process 34
Skorohod J_1 topology 15, 164
Spectral radius 108
Standard

Poisson process 34
  Wiener process 16
State space 78
Stationary
  independent jump intensity 96
  initial distribution for a markovian jump process 81
  random measure 43
  renewal process 43, 57
  transition probabilities 78
Steady state waiting time 126
Stieltjes transform 49
Stochastic approximation 31
Stochastic matrix 83
Stopping time 9

T
Top process 46
Traffic intensity 125
Transition probability 78
Two-state Markov process 48

V
Vague convergence 41
Vague topology 41
Virtual waiting time 126

W
Waiting time 126
Watanabe's theorem 38
Wiener process 16